Modelling and Simulation in Air Traffic Management

Springer
Berlin
Heidelberg
New York
Barcelona
Budapest
Hong Kong
London
Milan
Paris
Santa Clara
Singapore
Tokyo

Lucio Bianco - Paolo Dell'Olmo
Amedeo R. Odoni (Eds.)

Modelling and Simulation in Air Traffic Management

With 44 Figures
and 26 Tables

Springer

Professor Lucio Bianco
University of Rome "Tor Vergata"
Dipartimento di Informatica, Sistemi e Produzione
and Consiglio Nazionale delle Ricerche
Progetto Finalizzato Trasporti
Viale dell' Università 11
00185 Rome, Italy

Dr. Paolo Dell'Olmo
University of Rome "Tor Vergata"
Dipartimento di Informatica, Sistemi e Produzione
Via della Ricerca Scientifica
00133 Rome, Italy

Professor Amedeo R. Odoni
Massachusetts Institute of Technology
Department of Aeronautics and Astronautics
Cambridge, MA 02139, USA

ISBN 3-540-63093-7 Springer-Verlag Berlin Heidelberg New York

Cataloging-in-Publication Data applied for
Die Deutsche Bibliothek – CIP-Einheitsaufnahme
Modelling and simulation in air traffic management: with 26 tables / Lucio Bianco (ed.). – Berlin; Heidelberg; New York; Barcelona; Budapest; Hong Kong; London; Milan; Paris; Santa Clara; Singapore; Tokyo: Springer, 1997
ISBN 3-540-63093-7

Printed in Germany

Hardcover-Design: Erich Kirchner, Heidelberg

SPIN 10547486 42/2202-5 4 3 2 1 0 – Printed on acid-free paper

FOREWORD

This volume is a compendium of papers presented during the International Workshop on Modelling and Simulation in Air Traffic Management (ATM) that took place in Capri, Italy on October 2-6, 1995. The workshop was organized by the Progetto Finalizzato Trasporti of the Italian National Research Council in cooperation with the Massachusetts Institute of Technology (MIT) and the Federal Aviation Administration (FAA).

This was the fourth in a series of meetings held periodically over a ten year span for the purpose of encouraging an exchange of views and findings by scientists in the field of ATM.

The papers presented at the workshop dealt with a wide range of topics and covered different aspects which are currently important in ATM. However, all the papers contained in the volume were selected solely from that set of papers that addressed the main area of emphasis in the workshop, namely mathematical models and computational algorithms for ATM systems.

The development of these tools has obviously benefitted greatly from the evolution of digital computers and the enormous increase of available computing power. At the same time, the rapid growth of air traffic volumes in most of the world has increased the complexity of the air traffic system and consequently the complexity of the models used for the representation and analysis of ATM systems.

The papers selected attempt to depict the most up to date state of the art in this direction.

In the first paper of the volume, A. Zellweger provides a general introduction on the impact of the changes we have witnessed over the past two decades in the technologies utilized in the ATM field. This technology revolution, it is argued, has created tremendous opportunities as well as new challenges for ATM. In the next paper, R.W. Simpson describes a framework for analyzing new operational concepts for automated ATM systems. In particular, the author underlines that the new concepts allow compatible operations of aircraft with a broad range of capabilities and provide much of the significant increase in capacity which is needed in certain high density traffic areas.

The expected evolution of traffic characteristics and its consequences are considered in the third paper by M. Pelegrin. Three major factors are taken into account: the possibility of a fully automated flight, the use of an unique navigation and landing system and the concept of "free flight". In this context a global optimization in a large region is analyzed and discussed.

The paper by G. Andreatta, L. Brunetta and G. Guastalla deals with the ground-holding problem, i.e., the problem of reducing congestion at major airports by imposing a ground delay on a set of selected aircraft. For this problem many mathematical models have been proposed but they have found little application in practice, probably because of their complexity. To overcome this difficulty the authors introduce and analyze a family of simple heuristics, each one based an a specific "priority rule". The performance of these heuristics is compared with that of an "optimal strategy" on a set of test problems.

The paper by M. Lucertini, S. Smriglio and D. Telmon introduces a representation of a Flight Information Region as a service network and proposes the utilization of "pull" type strategies, derived from the manufacturing environment, in real-time flow management. To illustrate the effect of these pull strategies some computational experiments are carried out.

Next, D.G. Denery and H. Erzberger describes CTAS, a system for the automated management and control of operations in terminal areas, developed at NASA Ames in cooperation with the FAA. In particular they discuss the role of simulation and field experiments and present the most recent results of CTAS - related work.

The problem of real-time control of the Terminal Area is considered by L. Bianco, P. Dell'Olmo and S. Giordani. In particular they propose scheduling models and solution algorithms for the static and dynamic cases. For both cases, computational results are provided on a set of data similar to real ones.

Some human factor aspects in ATM systems are considered in the paper by R.J. Hansman, J.K. Kuchar and E.N. Johnson. They suggest an integrated "human centered" systems approach which considers the human controller as a functional component of the closed loop information system.

The last paper by J.E. Evans focuses on assessing how much of the $ 3B (US) per year in adverse terminal weather delays is "avoidable" and which ATM investment options would be most effective in eliminating the delay. The author proposes in particular two conceptual models for delay and considers various options for reducing the delay in the context of these models. He also makes suggestions for better assessing the potential benefits of the various ATM options.

In conclusion, this book is by no means a complete survey of the numerous ongoing research efforts on ATM systems. However, the papers included in the volume and the other papers presented during the workshop in Capri provide further evidence of the continuous progress being made in ATM system development.

The editors hope that this volume will help stimulate further research in this field.

Lucio Bianco **Paolo Dell'Olmo** **Amedeo R. Odoni**

Acknowledgement

We wish to thank F. Casavecchia for his contribution to organizing and conducting the workshop in Capri on which this volume is based.

We also gratefully acknowledge C. Landi for her careful assistance in the editorial work.

CONTENTS

TECHNOLOGY EVOLUTION AND ITS IMPACT ON AIR TRAFFIC MANAGEMENT

Andres G. Zellweger

Director of Aviation Research
Federal Aviation Administration
800 Independence Avenue, Washington, USA

In this paper, an attempt is made to complement the significant advances in Air Traffic Management (ATM) by providing a somewhat different perspective. The aim is to reflect on the impact of the changes due to the application of new technologies we have seen over the past two decades that provide the foundation for ATM. The thesis is that the technology of the past two decades presents an entirely different class of problems to be solved by today's engineers and scientists engaged in furthering the state of the art in ATM and also presents a radically new paradigm for solving these problems.

1. Introduction

The continued projection of growth in aviation as well as airline and civil aviation economics demand changes in Air Traffic Management. In the United States, the Air Transport Association claims that delays attributable to the Air Traffic Management (ATM) system cost the U.S. airlines billions of dollars per year. While air transport is the safest mode of transportation (last year, in the United States, more people were killed in animal drawn vehicle accidents than in commercial aviation), the flying public cries out for a safer system. Developing nations are finding that aviation is a promising underpinning for the transportation infrastructure necessary for economic development. Modern technology and innovative ways of using this technology have the promise of unprecedented aviation system advances necessary to meet the economic challenges, to further the potential of air transport, and to improve the safety of the aviation system in countries ranging from the U.S. with many thousands of aircraft and millions of operations per year to small nations, such as Fiji, with only tens of aircraft.

The papers that have been assembled for this workshop reflect the significant advances in air traffic management. They describe technologies and the application of technologies that are truly revolutionary. In this paper, I will attempt to complement them by providing a somewhat different perspective. I would like to reflect on the impact of the changes we have seen over the past two decades in the technologies that provide the foundation for Air Traffic Management. My thesis is that the technology revolution of the past two decades presents an entirely different class of problems to be solved by today's engineers and scientists engaged in furthering the state of the art in ATM and also presents a radically new paradigm for solving these problems.

2. Solving Different Problems

Some twenty odd years ago, when I began my career in aviation, the air traffic management community was pushing the limits of technology. The Air Traffic Management computers, software, displays, as well as the communications, navigation, and surveillance systems we were using were at the leading edge of technology. Commercial products were not suitable for our applications. In essence, we and the various defense agencies of the developed countries were the sole users of many of the technologies and, in those cases where we were not, we certainly had more stringent demands than the rest of the users of these technologies. We were in what is best described as a technology-pull posture - that is, we were pulling the technology advances in the direction of our needs.

Today, the situation is very different indeed. The ATM community is no longer the driving force for most of the technology being used. In the computer hardware and software area, in communications, and even in navigation, we represent a very small segment of the market and, in most cases, the capability of the technologies in these areas are more than adequate for Air Traffic Management needs. My lap-top computer, which is so small that I can carry it in my briefcase, is more powerful, has more memory, and a lot more software that most computers used for radar approach control in the United States! In Japan, more Global Positioning System (GPS) receivers will soon, if not already, find their way into automobiles each year than we envision as the entire market for aviation.

The question that faces decision makers is no longer "can we get a technology to meet our needs?" rather, it is "How do we best use the available technologies to meet the needs of our customers?". There has been a shift from R&D to get the right technologies for ATM to R&D on how to use technology and to decide what to implement. Today, technology is pushing ATM into new uses. The challenge being faced by our engineers and scientists is no longer the development of technology to meet our needs, rather, it is the determination of which technologies to use and, more importantly, the exploration of how to use the available technologies most effectively. Certainly, there are some cases where available technology does not meet unique ATM requirements, but, as I will argue later, economics is increasingly making us take a very hard look at our requirements and the way we do ATM to see whether we can use available technology by changing the way we conduct our business.

Just as there have been major advances in technology, the aviation system itself has changed over the past twenty years. Today we have a truly global system, no longer a set of loosely coupled ATM systems that can operate relatively independently. This problem has been particularly constraining in Europe, and has given rise to the EATCHIP program (European ATC Harmonization and Integration Program) aimed at providing common, integrated systems across Europe. The ICAO Future Air Navigation System (FANS) concept has provided a foundation for the global ATM system for the 21st century and is being used as the framework for the regional ICAO ATM/CNS Implementation Plans world-wide.

To some extent, this has been the result of the technological revolution, but for the most part, the need for a global system has come from a change in the underlying transportation system. Air transport has expanded tremendously, both in volume and in the areas of the world that it serves. The habits of the flying public, both in personal and business travel, have changed. Aircraft are larger, faster, with greater and greater range. Flight management systems and other avionics and aircraft systems have evolved to change the modern commercial airliner from an aircraft flown by its pilot to a complex system that is managed by a crew working together and interacting constantly with the airline operations centers and ATM systems on the ground. Airports are larger and busier, with a new set of problems that are often exacerbated by the public's growing environmental concerns.

To accommodate this change, the ATM system that we build for the next century requires us to deal with a new set of problems. We will have to deal with ever increasing complexity, we will have to accommodate the global, seamless requirements of the transportation system, and we will have to accommodate the growth in aviation in a manner that continues to make aviation an economically viable mode of transportation. In the U.S., as in many other countries, we must also provide for the needs of military and general aviation participants.

At the same time, even though air travel is by far the safest form of transportation, we will have to increase the level of safety because the flying public would not tolerate the increase in number of accidents that would occur at today's level of safety as traffic grows in the future. The FAA, for example, has, within the past year, set a goal of "zero accidents" and has put in place a senior safety official, reporting directly to the Administrator, to help us understand what needs to be done to increase the level of safety of our aviation system.

The increase in complexity will be a major driver in the type of ATM system that is needed. Twenty years ago, individual aircraft were handled by individual, rather loosely coupled sectors. Interactions among ground system components was limited to hand-offs between sectors. Today, we see ever increasing interaction between the ground (both the ATM system and the airline operations centers) and the aircraft. Data link is a reality and offers new opportunities. As demand continues to increase, capacity concerns must be addressed and it is becoming more and more important to manage the flow of traffic to optimize scarce resources. This in turn leads to interaction among airport towers, approach control facilities, en route centers, and flow management units. The conflict probes go across many sectors and even across ATM facility boundaries. To achieve the efficiencies dictated by airline economics, airlines will no longer optimize their operation independently of the CAAs. It will be necessary for airlines and CAAs to exchange data and make joint decisions about the flow of traffic. But the largest impact is likely to come from the increased capability of the modern aircraft and the potential change in the allocation of responsibility between the system in the air (pilot and avionics) and the system on the ground (controller and ATM decision aids). The Eurocontrol PHARE Project has, for example, developed an experimental system where pilot, flight management system, ground computer, and controller negotiate flight trajectories via data link and where controllers use new decision aids to reduce workload and thereby increase capacity.

Collaboration will be the norm in developing the future ATM system. ATM system design decisions will no longer be made alone by individual Civil Aviation Authorities. The nations of the world will have to plan jointly. There will also be much more active participation by the customer - the pilots and the airlines - both in terms of the role they play in the future system and in terms of the equipment they buy for their aircraft and their airline operations centers. I believe this collaboration to be an essential ingredient in making possible many of the new ATM methods that will be discussed in this symposium, in allowing the effective use of airspace across national boundaries, and in the cooperative management of airline operations and the ATM system.

In summary, we have had a revolution in computers, communications, and other technologies that will have a far reaching effect on the ATM system. There is a need for: a more global system, characterized by greater demand for service; increased safety; and methods for accommodating the increased complexity and greater interaction among the many elements. We have a different set of equations that governs what needs to be implemented and what R&D is needed to make possible the ATM system for the next century than we had twenty years ago.

3. Solving Problems Differently

In this section, I will address the paradigm shift in system design and development that is taking place to accommodate the changing needs and new system complexities discussed above. My focus will be on how problems are solved in this new environment, with special emphasis on the opportunities offered by modern technology.

3.1 Starting with a VISION and a ROADMAP

There is an old adage that says that if you don't know where you are going, any road will get you there. If anything, the developers of ATM systems have taken this adage too seriously in the past. They have defined, in great detail, the state of an ATM system as it might exist in 10 or even 20 years and, on the basis of this, have produced rigid specifications for systems that fit this characterization. The end result has been less than satisfactory because often, when a system was finally implemented, many years after the specification was developed, the world had changed in ways that were not anticipated and the system did not match the need. This has been characterized as "managing our system development to the past" since the bases of system development and project management were the detailed specifications and plans developed at the outset of projects.

A better strategy, and one being adopted by many developers of systems designed for a long system life in a changing environment, is to develop a vision, at a fairly high level of abstraction, of what the ATM world might look like in 10 or 20 years. It would take into account a variety of factors, including expected technology evolution, commercial trends, and aviation needs due to projected demand and economic considerations. Based on this vision, a general system architecture is defined that provides what city planners call their master plan. Just as important is a set of standards and development guidelines for building systems that are consistent with the architecture - what the city planners call their building codes and zoning laws. Finally, one needs a roadmap that shows how the ATM system will evolve from its current state to that described in the vision.

The vision and architecture tell us what kinds of systems need to be built. The roadmap presents the timetable and system development framework. While system elements are built to a specification developed at the outset, these specifications include requirements to build according to the standards, within the framework of the architecture, and with mechanisms for pre-planned product improvements. This accommodates the expectation that the systems will change over time to meet changing needs. Such system evolution is generally in the direction pointed to by the roadmap, but, as new technologies develop in ways that were not fully anticipated or as the needs for the system change, the roadmap may take slight turns to accommodate the changes. The vision will have to be updated periodically, as dictated by the new technologies and needs. This strategy will only be as good as the building codes and zoning laws we establish thus we must focus our energies on getting as robust a vision and architecture as possible. In essence, this paradigm for system development represents a shift in "managing to the past" to "managing to the future" (building codes and zoning laws that guide the development of evolutionary systems consistent with a vision of the future).

3.2 Developing Operations Concepts

Cockpits designed in the 1960's and 1970's were notoriously poor in allowing the pilot to carry out his or her job in a safe and effective manner. Similarly, the introduction of new features into the ATC system have at times failed miserably. The attempts to introduce a wake vortex advisory system in the U.S. in the 1980's failed, among other reasons, because controllers and pilots found it to be operationally unsuitable. The lesson was that, before a new element of an aircraft or ATM system is specified and designed, an Operations Concept (usually called an "ops concept") must be developed. To understand how, for example, a modern cockpit should be designed, one must first understand how the people (in this case the cockpit crew) and the machine (in this case the electronic and physical design of the cockpit) will work together to carry out the proposed mission.

In the past, ops concepts have been developed by performing some sort of mission analysis and then having a team of "users", engineers, and human factors specialists develop a paper and pencil product. While this was initially satisfactory, designers of both aircraft and of ATM system elements are finding that as systems are becoming increasingly complex and interdependent, one can no longer predict the effects of the many ways in which pilots and controllers interact with one another and with their decision support systems.

One promising solution appears to lie in the use of distributed interactive simulation made possible by the technology of the 1990's. Scientists can develop simulations that represent different interacting components of a proposed new system, as well as related systems, at various levels of abstraction, depending on the specific problem at hand. Experiments can then be run with people in the loop exploring realistic alternative ops concepts.

A simple example that has led to the successful implementation of a new procedure to save fuel in trans-Pacific flights illustrates the idea. About two years ago, a group of pilots were complaining to controllers that in-trail restrictions often prohibited an aircraft from climbing to an optimal fuel altitude because it was not permitted to climb through an altitude occupied by a flight in front of it if the two aircraft would get closer than the required separation distance during the maneuver. A proposal was made that if the pilots of the two aircraft involved could establish radio contact, if the pilot of the trailing aircraft could establish positive identification of the leading aircraft on his or her TCAS display, and if a climb maneuver could be coordinated with the oceanic controller, a safe climb could be executed. Initially, there were numerous questions about the viability of such a maneuver, but within two or three weeks, an experiment involving a simple cockpit simulator with a TCAS display and a console representing the oceanic automation system was developed in an FAA laboratory (called the I-Lab) at Mitre to demonstrate the ops concept to both pilots and controllers. It was quickly determined that, with some minor changes in the ops concept, there was sufficient promise to warrant further research. Subsequent sections will show how, using more detailed simulation and prototyping technology, the in-trail climb maneuver was implemented less than two years after the initial concept was postulated. While this example does not represent the complexity of most of the new ops concepts, it serves to illustrate the idea and is proof that such interactive simulation can serve an important role in ops concept development and can lead to accelerated implementation of new concepts.

System level simulation during ops concept development and refinement allows determination of stability of the proposed human-machine system. Heavy loads, system anomalies, and failures can be introduced to determine whether the ops concept is robust under a wide range of potential situations. The failure recovery mechanisms can be validated with humans in the loop. The system level simulations answer critical questions about proper allocation of functions to humans and to computers and, perhaps most important, give initial insight into the inherent safety of a new way for machines and people on the ground and in the air to work together.

Ops concept development with the use of system level simulation generally provides the foundation for building new systems or for providing significant enhancements to existing systems. Another use is in defining new policies for Air Traffic Management, as seen in the following example. The Chicago convention makes the controller responsible for separating aircraft from one another and from the ground; responsibility for separation of aircraft from weather phenomena is strictly in the pilot's domain. But since the Convention was established some 50 years ago, aircraft performance, conditions under which aircraft fly, and the technology for detecting and predicting weather phenomena have undergone tremendous change. In the past, pilots were able to make decisions about what to do in the face of adverse weather by looking out the window. It is clear that this is no longer adequate and that the policy regarding the controller's role in separating aircraft from weather must change, but just what a new policy should be is not so clear. In the past, a policy change would have been formulated by careful deliberation by the affected stakeholders with the support of technologists. This worked so long as systems were simple and if the role of the human did not change significantly. For the weather example that is not the case. System level simulation that explores alternative technology implementations (different mixes of the types of weather information that can be made available to pilots and controllers) and different alternatives for new responsibilities for pilots and controllers can make a significant contribution to determining what new weather delivery and presentation systems should be used and what the corresponding policy for pilot/controller responsibilities should be.

3.3 The Contribution of Prototyping

Simulation to validate an "ops concept" proves the feasibility of a new concept and is a necessary step in ensuring success, but, as we all know, the devil lies in the details. A great deal of work has to done to take a concept to successful fielded implementation. Technology advances, in both the hardware (computers, display systems, communications, networks, etc.) and in software (code generation, re-use, commercial products for quickly building computer-human interfaces, etc.) have put prototyping into the center of the system development paradigm. Simply stated, a prototype is an early implementation of a system where only those portions of a system that are critical to a particular study are represented, at whatever level of fidelity is necessary to trust the results of experiments. In essence, one is creating a virtual environment of a future ATM system for the purpose of studying various aspects of that new system.

Usually, after a concept is shown to have promise, we move into a research phase. Fairly simple prototypes are built to explore specific research issues. As a research project matures we see quite sophisticated prototypes that are used to simulate a substantive part of the system. For example, the R&D prototype used at NASA Ames to develop the Center Terminal Automation System (CTAS - a set of decision support tools to help the controller manage traffic flows in congested terminal areas), contained realistic representations of en route and terminal sectors. Airspace for a particular geographic area (Denver) was represented and controllers from Denver came to NASA to participate in design and evaluation experiments. These experiments served to refine and validate both the underlying decision support algorithms as well as the ideas that had been proposed for the controller-computer interface. Later experiments added cockpit simulators to the system to better understand how pilots operate in this new environment and to understand how data links could best be used for more advanced versions. This type of prototyping allows us to conduct repeatable, scientific experiments in a closed-loop environment, but it does not always predict how well a system might operate in the "real world".

I recently read an editorial in a consumer magazine for sailing products in which the author was lamenting that whenever they tested a product in a laboratory, under rigorous and controlled conditions, readers would complain that labs could never duplicate the multitude of conditions in the real world. When they tested a product under real conditions, through sea trials in a range of conditions, readers complained that results couldn't be trusted because the experiments were not conducted in a scientific manner, under controlled conditions. As a result, when the engineers at the magazine have the budget and the time, they test new products in both environments. We have a very similar situation in the ATM development arena.

While early efforts at prototyping ATM improvements were limited to the laboratory, technology advances now make it possible to go well beyond that laboratory and to conduct experiments under much more realistic conditions. The first step, in the CTAS example at NASA Ames, was to tap into the ATC computers in Denver and to use telecommunications to bring live data into the experiments at Ames. Later, the prototypes were moved to an ATC facility in Denver, were connected to all the appropriate data sources, and run in "shadow" mode. (In the case of simpler systems, such as the data linking of departure clearance introduced in the U.S. a few years ago, prototypes are sometimes brought to field facilities for full scale operational tests.) While field experiments using prototypes do not usually permit the closed loop simulations that are necessary to fully understand the behavior of a new system under a variety of conditions (high workload, failure scenarios, etc.) a great deal is learned when one can have both scientists and the controllers at a facility evaluate how the new system behaves, in a passive way, under the real world conditions. As the boating enthusiasts realized, we often don't know enough about how the real world parameters interact with a complex system and don't understand the real world parameters well enough to simulate them accurately in the laboratory. In essence, we now have the best of both worlds - our challenge is to understand our problems well enough to define the proper fidelity for experiments and to achieve the right balance of laboratory and field experiments.

Prototyping of system components works well to study a new system element in isolation, but, as I have argued earlier, the ATM systems that we expect to have in place in the 21st century have much more interaction among the many system elements. To explore these in the laboratory today, we can use distributed simulation technology to connect prototypes of different components under development. For example, a CTAS prototype was connected to and Automated En Route ATC (AERA) prototype and to a traffic flow management prototype in an FAA lab to explore the complex interactions between the computers systems and the people in a future environment. Today's technology allows researchers to connect the ATM laboratories in a number of locations and to also tie in high fidelity cockpit simulators to explore how all parts of a future system work together under new ATM regimes. As in the following example, scientists even bring real aircraft with prototype avionics into a simulation when they believe it is critical to understand aircraft and pilot performance and behavior in a real world environment. Analysis of how weak or how strong the interdependence and interactions of various system components is determines the required fidelity and types of system level simulations that are necessary during the research and development phases of a project.

Prototyping not only leads to the design of better ATM system components that are more suited to the real world environment when implemented, but it serves to ensure user acceptance and to reduce the time it takes to move from the inception of an idea to its operational use. The In Trail Climb (ITC) example introduced earlier is a case in point. After the initial concept was accepted, FAA, together with airlines and pilots, continued prototyping activities at its Technical Center in Atlantic City, this time adding much more real world fidelity to the simulation. An Oceanic Laboratory that is a replica of the system used to control oceanic traffic in the U.S. and high fidelity cockpit simulators were used. Later, live flights were added into the simulation. The final step consisted of real world trials of the ITC maneuver over the Pacific. My own estimate is that time from concept to successful use was cut in half because we had the technology for the different stages of prototyping of ITC.

As this simple example showed, prototyping lets us explore how to best implement a new concept or how to build a system during all phases of system development - from early research to detailed design. It allows us to address and resolve many of the major risks in the early stages of a program. In essence, we can study, how a proposed system of people and machines will behave before we have to make firm design decisions that could be very difficult and costly to change later. The complexity of ATM systems today requires prototyping because we simply cannot afford to build and implement a system based on paper designs and expect that it will work as intended!

If anything, prototyping has been too successful. Often, the controllers and pilots who participate in real world experiments become so enthusiastic about a new system and the benefits they envision from use of the system, that they want immediate implementation. Unfortunately, prototypes usually don't contain all of the features necessary for full implementation. The software in a prototype is almost never suitable for use in a production environment, it does not have the inherent robustness nor does it have the maintainability characteristics required in a fielded system. The developers must face the difficult job of managing expectations and research is needed to find a new system development/production paradigm that accelerates the process of moving from the prototype stage to full implementation.

3.4 Changes in Building Systems

Both advances in technology and the changing character of the ATM systems we are building today have an impact on ATM system acquisition. Technology is helping us solve some of our traditional problems and, at the same time, technology advances and the systems they make possible, are leading to a new set of problems that require a solution. In this section, I will explore six different trends and their implications: the separation of the hardware and software life cycles; the increasing desire to use commercial, off the shelf components, especially software (COTS); the advances in development of the graphic user interfaces (GUI); the growing concern with security; the increasing difficulty of ensuring system integrity; and, finally, an increased need to deal with the "human element" through application of human factors principles.

ATM automation systems of 10 or 20 years ago were integrated systems of hardware and software. We considered them as a single system and, when replacements were necessary, new hardware and software came as a single package. Today, this paradigm no longer works. The hardware and software life cycles are different and we must produce new ATM systems with full recognition of the implications of this separation of life cycles.

As I argued earlier, ATM has become a user of commercial computer hardware. Because the advances in hardware are so rapid, hardware generations last only two to five years. It is more cost effective to replace hardware as new generations are introduced than it is to keep old hardware and pay higher maintenance costs. I generally replace my personal computer every two to three years. The challenge is to build our embedded systems to accommodate this life cycle strategy for computers. Modern controller workstation must, for example, be built in a way that permits replacement of computers and other electronic components on a periodic basis.

ATM software, on the other hand, is getting more and more complex and voluminous. As has been witnessed by the problems faced by builders of large ATM software systems worldwide, it takes a long time to develop a new generation of software. That means that software life cycles are still on the order of 20 years. To accommodate the evolutionary nature of ATM that I described earlier, it becomes necessary to have software that can continually be enhanced over this long life cycle. In effect, the paradigm must change from one of software maintenance to one of continual software upgrade. New software releases are driven by the need for new functionality, not by necessary fixes to old problems. While open system concepts and development of a system architecture show some promise, the need for software systems that can evolve is pushing the state of the art in software engineering.

The difficulty and cost of developing large ATM software systems is leading ATM system developers to explore the use of readily available commercial software components. Clearly, the use of COTS can be advantageous, but with it come some problems as well. COTS is driven by the commercial market and tends to have a much shorter life cycle than ATM software, thus one must devise a strategy for either replacing COTS periodically or else maintaining COTS long after it has been abandoned by its original developer. There are also concerns about COTS failures since the reliability/availability requirements of ATM systems is significantly greater than that for most COTS. Finally, COTS is not designed specifically for ATM and thus doesn't necessarily meet all of the requirements for ATM. This requires judicious review of our requirements and a determination of what software features are absolutely essential. Some have suggested that one should explore how to do the ATM job differently to take advantage of COTS. To make the best use of COTS in ATM we must select the right standards, have good insight into where commercial technology, as well as standards, are moving, and finally, continue testing proposed new developments against our architecture.

Traditionally, one of the most difficult tasks in building ATM software was the design of the graphic user interface (GUI). We had to do the R&D for a GUI early in the development of a new system because the GUI design had to an integral part of the software. Once the software was designed and built, changing the GUI was quite difficult. Today, software technology for GUI implementation has developed to the point where details of the interface (data presentation formats, colors, symbology, location of data, etc.) do not need to be completely understood and finalized before software is built. Tools, called GUI builders, can automatically generate the code for the GUI and integrate it into the rest of the software. Thus, instead of difficult research long before a system goes to use, today, we build a system with a rough approximation of a GUI that we think is likely to work. The system is then taken to a field site and the users experiment with the system for several months to finalize the most suitable user interface. This has revolutionized how we build ATM systems, but there are some cautions. While the "front of the panel" GUI can be changed quite easily during the life of the system, it is still very important to predict the evolution of the ATM functionality to make sure that the underlying functions that support the computer human interface can be adapted during the projected system life.

As we have seen, advances in technology have made the building of ATM systems far easier, but they also bring with them new challenges for system developers. The greater interconnection of computers through networks, the use of COTS, and the increase in computer literacy in the general public bring with them a much greater security risk than we had in the early days of ATM computing. This will necessitate safeguards both in building and in the use of modern ATM computing systems. We will need firewalls to protect the systems from viruses and malicious intrusion by unauthorized software engineers.

Another set of problems comes from the simple fact that, as we increase the sophistication of the decision aids provided to pilots and controllers, reliability of the ATM system needs to increase. One can no longer depend strictly on the human to back up a system when it fails. The FAA, in its design of ATM systems, such as the new Voice Communication and Switching System, has established a policy that there will be a dissimilar, albeit less capable, backup system. No matter how high the specified level of availability and reliability, it is our assessment that with the current state of the art in software development, we must have independent backup systems - software will fail and we must be able to maintain safe air traffic management in the rare instance when that happens.

In some cases, maintaining system integrity becomes more critical because failures can affect much larger portions of the ATM system. A good example is in the move from the traditional instrument landing systems (ILS) to a GPS based landing system. Based on cost considerations, the U.S. has elected to provide the augmentation necessary for a GPS based Category I landing system through a wide area augmentation system (WAAS). Instead of independent Cat I ILS systems at individual airports, whose failure will have effects on only a single airport, failures in the WAAS can impact many airports across a wide region. Thus the design of a WAAS requires special considerations to guard against failures.

Some twenty odd years ago, I remember coming to a EUROCONTROL Seminar to talk about FAA's research into "automating" air traffic control. We thought that we could design computers that could do the job of the air traffic controller much better. The audience was at best tolerant, but I'm sure there were a lot of hidden laughs about this crazy young American scientist. Today, I've come full circle - I no longer believe that we can build systems with the requisite reliability to do ATM automatically, but, more important, I've learned that we strive, not for "automation", but for decision support tools that lead to a less labor intensive (we call it "more productive") and a more efficient ATM system by designing computers to help people do the things that they do well and at the same time to achieve the objectives of safety, productivity, and efficiency. We call this "human centered" automation. In the next two decades, we will see more and more decision support tools, many approaching the sophistication of the Pilot's Associate developed by the U.S. Advanced Research Project Agency. This system had sophisticated data fusion and used intelligent agents in its software to truly complement the pilot in ways that made his or her job easier and, at the same time maintained system safety. As we begin building systems like this for commercial aviation, the discipline of human factors will play an ever increasing role in ensuring that we truly "humanize" technology.

4. Conclusion

I hope that I have convinced you that the technology revolution has created tremendous opportunities as well as new challenges for Air Traffic Management. But let me give you some examples of our recent successes:

- New control centers designed only a few years ago have computer rooms that are standing empty because computers are so much smaller.

- ARPA spent millions of dollars to design a small GPS receiver. When the project was finished, a commercial receiver of identical size and capability could be purchased in a local store for a thousand dollars.

- Spectrum constraints have led us looking for more and more efficient use of the voice radio system. We have split the frequency once and there are plans in Europe to go from 25 khz to 8.33 khz channel spacing. The FAA is doing research on TDMA to get even more capacity. But, as we move to Free Flight, the need for voice communication in en route airspace diminishes and the delays of commercial satellite communication will probably be acceptable. Thus commercial communication technology, which has for years been rejected, may very well be the solution for at least some air ground voice communication.

- Until very recently, FAA was in the process of designing and building a microwave landing system. Millions of dollars had been invested when a decision was made to cancel the contracts because of the rapid progress in R&D for very precise landing systems based on GPS. Only a few years earlier, the accepted wisdom was that GPS might work for non-precision approaches but would certainly never have the precision necessary for CAT I, much less CAT II and III landing.

- I could go on with many more examples that show that we have not been very good at predicting the future. The technology revolution that I have talked about will continue. Just as I have argued that today we must solve new problems in new ways, in ten years we will be faced with another set of problems and opportunities. Our challenge will be to understand the future well enough so that we can make the right decisions to take advantage of the potential of new technology, to make the correct design and investment decisions. Our best thinkers must look to the future to ensure success.

CREATING NEW OPERATIONAL CONCEPTS FOR GLOBAL AUTOMATED ATM SYSTEMS

Robert W. Simpson

Professor of Aeronautics and Astronautics
Director, Flight Transportation Laboratory
Massachusetts Institute of Technology
Cambridge, Massachusetts 02139, USA

With the availability of new CNS (Communications, Navigation and Guidance, Surveillance), tools as described by the ICAO FANS agreement and the potential of new tools for Automation at the ATC controller's console, the aviation world is once again in the same position as it was in 1945 when radar and ground based radio-navigation aids were introduced to create the present forms of ATC operations. But instead of providing the tools and waiting to see what the operational people will create in the form of new ATC operating procedures with them, the size and global scope of the new investments make it necessary to do a proper, top down approach to engineering a new ATM system which brings real benefits to the aviation community. This means pre-determining the set of new Operational Concepts and the details of their procedures which are safe and economical, which allow compatible transitions under mixed old and new operations, and which provide a significant increase in capacity needed in certain high density traffic areas. This paper is concerned with establishing a framework for describing and analyzing new ATC Operational Concepts.

1. Introduction

The problem can be stated quite simply: *"Given the CNS tools as agreed to by the ICAO FANS activities, how do we design and introduce new ATC operational procedures which allow global transitions to more automated, higher capacity, and more efficient ATM environments?"* but it involves developing a significant new engineering capability that does not exist today; namely, the capability to design, analyze, test, and validate new and radically different ATC operational procedures. A proper "top-down" engineering approach to building new ATM systems requires considerable work to establish valid and acceptable operational requirements which then, *and only then*, allow the establishment of technical requirements for CNS and Automation tools.

The engineering problem has several external, real world factors which must be understood:

1. There is a large number of aircraft owners and operators who will be required to invest in new onboard equipment. They will do so *when and if* the operational benefits of the new procedures exceed the costs of owning and maintaining the new equipment.
2. There should be benefits to the various operators of ATM systems around the world in the form of increased productivity of ATC controllers, and reduced investment, maintenance, and operating costs of ATC equipment.
3. The benefits to aircraft owners and ATM operators vary with traffic density so that the timetable for introducing the new equipment and procedures at various locations around the world will vary during the next thirty years. There will be a long transition period.
4. To shorten this long transition period, it is desirable to be able to operate old and new procedures simultaneously in the same areas and locations. This is technically possible since the coverage of new CNS tools is global. It requires some ingenuity in designing new ATC concepts.
5. The new operational procedures will be *automated*, and must be *identical* around the world for the air and ground equipment to be compatible, and for international pilots to avoid dealing with similar but inconsistent procedures. It is desirable that the enroute procedures for Oceanic areas and for low traffic density continental areas of the world become identical.
6. While these new operational procedures must be identical, there can be similar but differing technical implementations by different equipment suppliers.

There are also a variety of institutional and legal problems for the international aviation community to solve which arise from the above factors of designing what can be called a "*Global Automated ATM System*". The world's aviation lawyers and politicians must be involved if a successful outcome is to be achieved - but the engineers must explain to them exactly what is needed in these areas. For example, it will be operationally feasible in the future that continental enroute ATM services can be provided from only one multinational ATM Center; or indeed, that there only be two redundant enroute ATM centers for the entire world. There are financial, legal, institutional, and political problems to be solved in achieving the efficiencies of such radical innovations. But the improved performance and cost savings of such possibilities are enormous, and efforts to accomplish such consolidations should be undertaken.

2. Developing a Framework for Analyzing Concepts for ATM Operations

The key element of any ATC operational concept for air traffic control (and indeed for any type of vehicular traffic control system) is the process by which the future trajectories of aircraft are generated. I shall denote this future trajectory (or Flight Plan, or ATC Clearance, or Intended Path) by the symbol F, and shall note that it can be specified in four dimensions (lateral, longitudinal, vertical, and time). It is often incompletely specified today, and in particular, may only be specified a short period into the future which I shall call the "Planning Horizon", denoted T_p. The future trajectory starts from the current position of the aircraft, but note that information about the past position of the aircraft, P, as gathered from a radar surveillance system is irrelevant, and any inference from such P information that it will continue its present trajectory can be dangerously wrong. ATC concepts always use F information, and the feasibility of new concepts depends critically on the quality and content of this F information.

Identifying F and T_p allows the formalization of a traditional method from past studies of classifying ATC system operational concepts in two dimensions:

1. ATC concepts can be classified as *Strategic* or *Tactical* depending on the size of the Planning Horizon, T_p.
2. ATC concepts can be classified as *Centralized* or *Distributed* depending on where the F for each aircraft is being generated.

A long period of planning creates a Strategic concept, while a concept with a short planning period is called Tactical. Obviously, there is a continuum in the T_p dimension, and it is possible that there can be a mixture of Strategic and Tactical elements. While pilots originate requests for a Flight Plan for trips, any traffic problems cause it to be modified by one (or more) of the centralized ground elements concerned with the flight. Again, there can be a mix of centralized and distributed activities in determining F in an ATC concept whenever some of the dimensions of F (such as speed or altitude rate) are left unspecified by the central ground element. Some current operating concepts will now be discussed to illustrate the use of these terms;

For example, the present *Procedural* operating concept for non-radar operations in Oceanic and low density continental areas is a Centralized concept which uses Strategic planning of the routes and altitudes for the F of the complete flight, but the central ground element then uses *Tactical* planning of longitudinal progress based on frequent position/altitude reports by radio. In these systems, Tactical implies 15 minute, or perhaps hourly position reports along the flight. The position reports include intention information in the form of the estimated time/altitude at the next reporting point based on planned airspeeds and estimated or forecast winds.

The current operating concept for areas where there is long range radar coverage may be called *Radar Monitoring* since it allows continual tactical monitoring of actual lateral and vertical positions which should conform to the strategic Flight Plan, with tactical interventions to correct any errors or blunders by the aircraft. It also allows tactical enroute changes of routings and altitudes which become desirable due to weather and actual (as distinguished from planned) traffic situations. Importantly, it allows a reversion to a Procedural operating concept in the event of a radar system failure in an area without redundant coverage. Failure cases often determine the real world feasibility of a concept for ATC operations.

The present *Radar Vectoring* concept for high traffic density Terminal Areas is a Centralized/Tactical system where the extent of the planning horizon for *F* is very short (less than a minute), and pilots may not know much about their *F*. Indeed, controllers do not have an explicit *F*, but rather may have a general pattern for *F* from which they intend to make a specific path for each aircraft depending on the actual progress of traffic. This traffic pattern becomes familiar to each pilot. The radar vectoring controller has become responsible for the guidance of each aircraft, with the pilot responsible for conformance not to *F*, but to guidance "vectors" (in the form of air speeds, pressure altitudes, and magnetic headings) as they are tactically issued by the controller.

Finally, the *VFR Flying* concept used by the slower aircraft of general aviation in the lower levels of airspace under visual weather and low traffic density conditions is an example of a Distributed/Tactical system. In this concept, each pilot generates his own F, and can be quite tactical in changing it as the flight proceeds. VFR aircraft do not have any information on the *F* for other aircraft, and must continuously monitor the airspace as best they can to obtain visual contact with a nearby aircraft, and then try to infer its intentions by observing its track, its orientation, and its altitudes. When VFR aircraft are in radio contact with various ground elements of the IFR ATC system, these elements may ask about intentions, or ask to be informed about imminent changes of altitude and heading. Around small airports, pilots may broadcast their progress and intentions on a common radio frequency as a traffic aid to other VFR pilots.

3. Traffic, Encounter Rates, and Interventions

At this point it is necessary to digress in order to define some ideas needed to describe traffic activities. What do we mean by the word "traffic" when we are talking about traffic control systems for aircraft? Let me propose a definition:

Traffic is the interference between vehicle trajectories when they "encounter" each other in a way that requires an "intervention" to modify the trajectories.

An *Intervention* is a traffic control correction which modifies planned trajectories, F, to accomplish two goals:

1. to ensure safety by avoiding collision, or
2. to avoid congestion by reducing arrival traffic flow rates.

An *Encounter* is the meeting of two (or more) aircraft which requires an "intervention". Using analytic "Traffic" or Encounter "Models" it is possible to estimate the average Encounter rate in any sector of traffic operation if we know some sector traffic parameters. Encounter rate in any area or sector is a function of three traffic parameters:

1. The square of the average traffic density (aircraft per cubic mile),
2. The linear dimensions of the Encounter Volume defined by the safety separations or required flow rate,
3. The relative encounter speeds averaged over the total mix of aircraft in the sector (which depends on the patterns of traffic flow in the sector)

It is possible to find encounter rates by simulation models, but in most cases it is simpler to use the analytical models.

The *Workload* in an ATC sector consists of "routine" workload elements (which occur in establishing communications, providing weather and other flight information) which occurs even if there is no other traffic and depends on the arrival flow rate, and "traffic" workload elements which are associated with the interventions. The workload rate thus depends on traffic arrival rate in the sector, and the Intervention Rate.

Obviously, the Intervention rate in a sector equals the Encounter rate so that sector traffic workloads depend on the traffic density squared, the size of separation and flow criteria, and the relative encounter speeds. As traffic density doubles, the intervention rate quadruples. As separation or flow criteria are halved, the intervention rate is halved. If the relative speeds between traffic pairs are halved, the intervention rate is halved (e.g. all aircraft flying in the same direction at the same speed has zero relative velocity and zero interventions).

Traffic intervention rates can be high in a high density sector of airspace which means that the probability of any aircraft having a "free" flight in the sector become low. The probability of achieving a "user preferred trajectory" in a high traffic rate sector are low since they are likely to be modified to achieve the safety and congestion goals of the air traffic system.

4. The Basic Functions of an ATC Concept

There are three basic Functions which *must* be executed by all ATC Concepts. They are *Conformance, Separation Assurance*, and *Congestion Management*. The manner is which they are implemented operationally will allow us to define quite distinct concepts of ATC systems. Each Function can have defined Processes, and each Process can have different Options. There is one other ATC Function called Flight Plan Data Management which will be present in the implementation of any ATC system but which does not provide a basis for distinguishing between different ATC operating concepts.

4.1 Conformance Management

Conformance Management is the ATC function which monitors and corrects aircraft guidance errors in following a pre-defined trajectory, *F*, in two, three, or four dimensions. It ensures that aircraft are following their pre-determined conflict-free traffic path with an expected degree of precision called the Conformance Criteria (which may be variable during the flight depending on the traffic situation). It has two main Processes called *Conformance Monitoring* and *Conformance Resolution*.

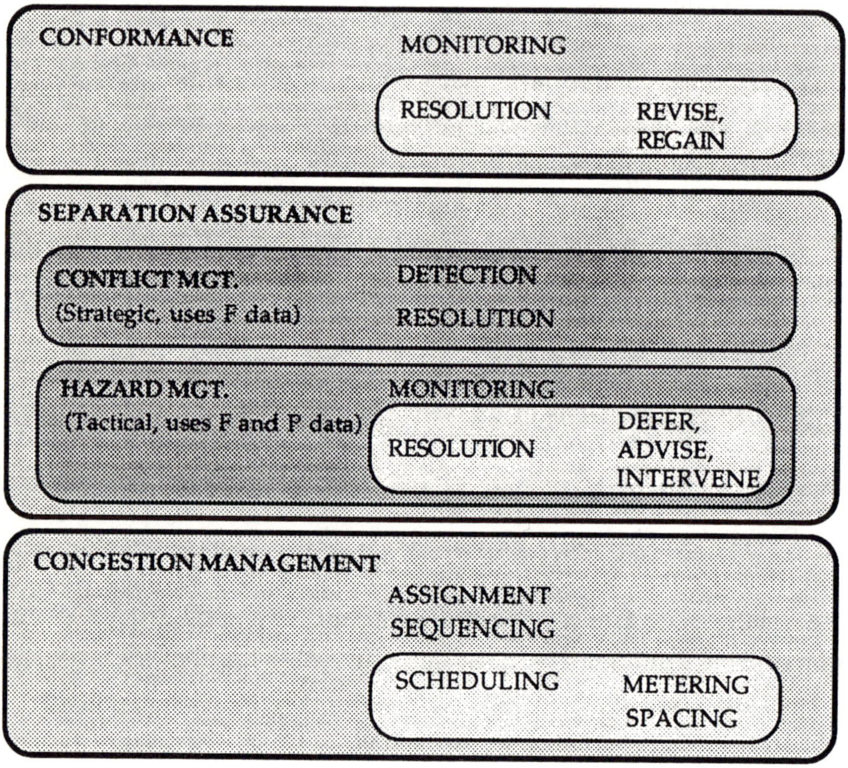

Figure 1. The Structure of ATC Functions, Processes and Options

4.2 Separation Assurance

Separation Assurance is the ATC function which ensures that aircraft have safe separations from each other and from terrain, and special use airspace. There are two types of Separation Assurance:

1. Strategic Separation Assurance which deals only with *F* information and which may be called *Conflict Management*. It has two main processes called *Conflict Detection* and *Conflict Resolution*.
2. Tactical Separation Assurance deals with both *F* and *P* information and which may be called *Hazard Management*. It has two main processes called *Hazard Monitoring* and *Hazard Resolution*. Hazard Resolution has three Options called *Defer*, *Advise*, and *Intervene*.

4.3 Congestion Management

Congestion Management (or Traffic Flow Management) is the ATC function which reduces flow rates to be less than the capacity flow rates of any ATC element. It has three main processes called *Assignment*, *Sequencing*, and *Scheduling*. The Scheduling process has Options called *Metering* and *Spacing*.

This paradigm for defining ATC functionality is illustrated in Figure 1, and the block diagram for a ground based sector showing the flow of information between these functions is shown in Figure 2 (taken from [1]). It is claimed that all concepts for ATC operations can be described by the functionality of this paradigm and the flow of information in the concept by the associated block diagram. Conversely, the paradigm can be used to explore the universe of ATC operational concepts. It is useful in defining the functionality of any proposed automation for an ATC system, and in studying the Human Factors of ATC controllers.

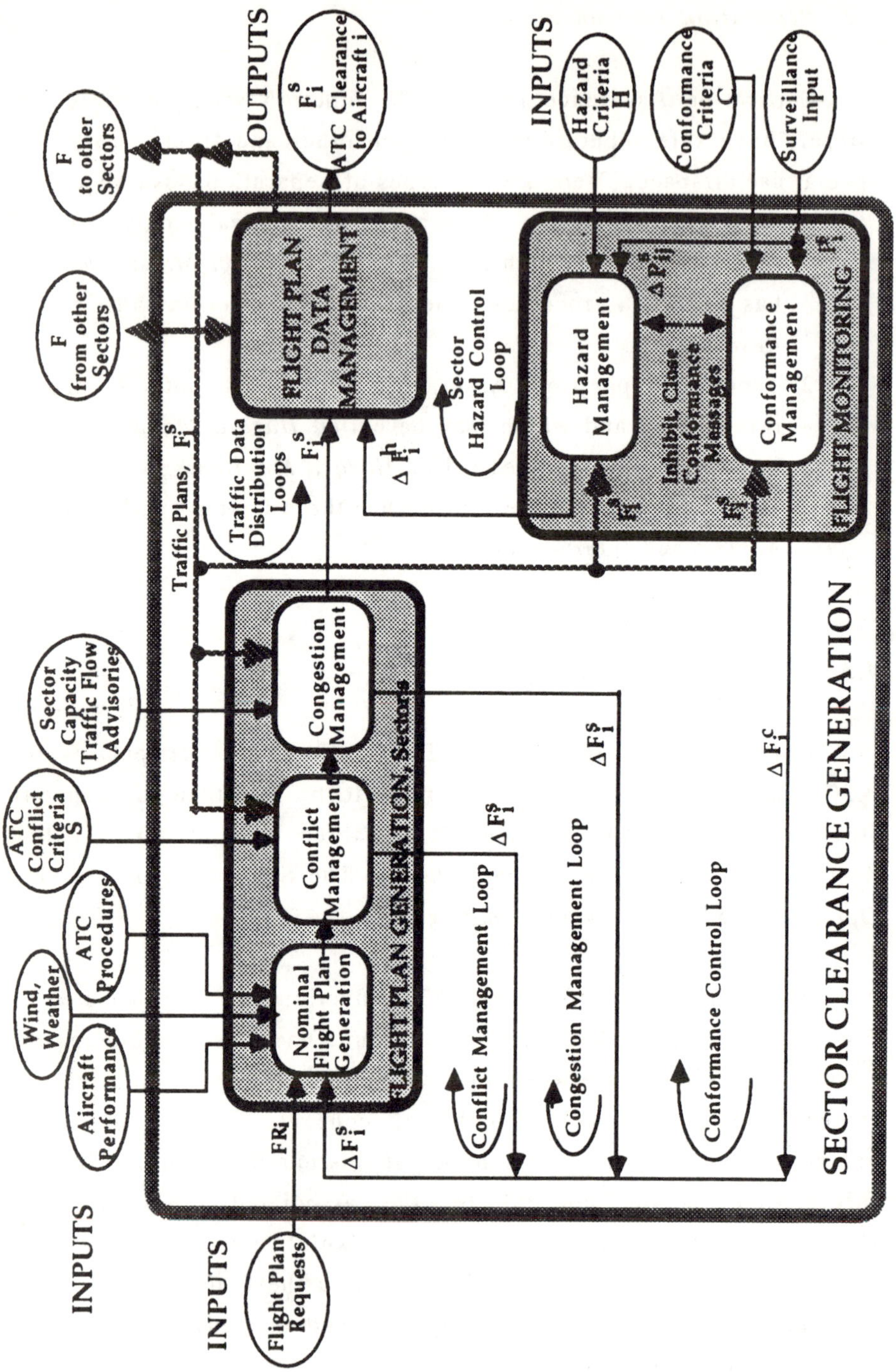

Figure 2. Block Diagram of Information Flow for ATC Functions and Processes

5. Goals and Issues in Developing an ATC Concept

There is a triple tradeoff in the engineering of ATC systems. While *safety* is paramount, and absolute Target Levels of Safety (TLS) have been established for a few ATC operational activities, there are also goals in terms of *capacity* and *cost*. To increase the capacity of any ATC element, there must be a decrease in separation criteria which implies a reduction in safety unless it can be shown to meet a desired TLS through improved reliability and performance of the ATC system in Conformance Management and Separation Assurance. Efficient use of capacity means a smaller investment in ATC facilities and equipment, but to get the required capacity there may still be a need for increased investments. Finally the cost of operations for aircraft can be reduced by procedures which provide less safety and less control over congestion and which require more ATC personnel. The Holy Triad of the goals for any ATC system concept are shown in Figure 3.

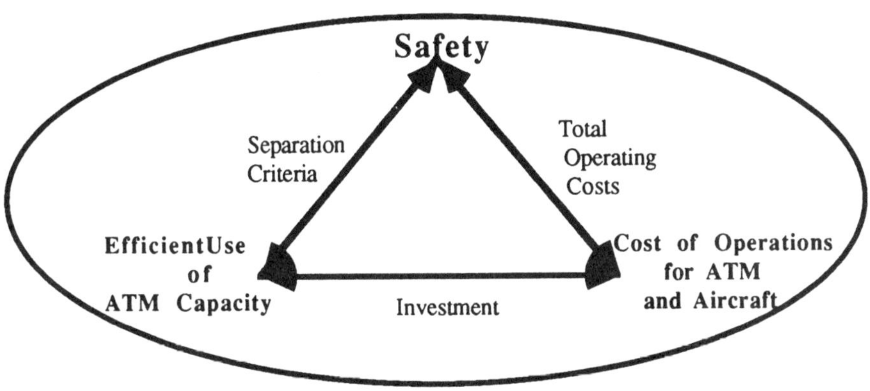

Figure 3. The Holy Triad of Goals for ATM Concepts Development

There are a very large number of issues in ATC operational concept development. The following is an incomplete list:

1. Who generates/modifies the Traffic Clearances?
2. What is the Planning Horizon for Traffic Clearances?
3. How often are Traffic Clearances updated?
4. What is the precision and dimensionality of the Traffic Clearances?
5. How and When is traffic intent information distributed to all participants?
6. What are the Conformance capabilities (accuracy and reliability) for aircraft?
7. How accurately are weather and winds forecasted over the Planning Horizon?
8. Are there Decision Support tools available for the ATC functions?
9. Are new ATC operational procedures compatible with existing procedures?
10. Do the new procedures have more capacity? Efficiency?
11. What are the complete investment and operating costs for the new concept?
12. What are the benefits of the new concept and when are they achieved?
13. How do we predict the safety of reduced ATC separation criteria?

6. Available New Technologies - The New ATC Toolset

Before describing examples of new concepts for ATC operations, it is necessary to review what proven technologies are now available for the designer of new ATC systems, and point out briefly exactly what new operational capabilities they offer;

Communications - The introduction of *datalink (DL)* technologies is probably the most important new step since it opens up the possibilities of automation of many ATC functions and processes. The new technologies offer three forms of two-way datalinks which can provide a capability for digital computers on the ground to exchange data with computers in the cockpit. Under the FANS ICAO agreement, it is planned that there will be free use of the *Mode S datalink (MDL)* within the coverage of the Mode S radar system in high density Terminal areas. For medium density enroute continental areas, there will be a moderate user charge per message to commercial service providers via a *VHF datalink (VDL)*. In Oceanic areas, there will be a *Satellite datalink (SDL)* where the user charge per message will be more expensive. Not approved by FANS, but trying to achieve acceptance, is the use of existing high frequency (HF) stations to provide datalink services *(HDL)*. Obviously, these systems will provide redundant overlapping coverage in certain areas. The aviation communications community is establishing a modern open systems protocol called the *Aeronautical Telecommunications Network (ATN)* as a basis for all future communications development, both on the ground and in the air.

Navigation - The introduction of *satellite navigation* offers a significant improvement (from miles to meters) in position measurement accuracy as well as global area navigation *(RNAV)* coverage over the complete aircraft movement from the start of taxi to landing and arrival at the gate. The initial form of Global Navigation Satellite System *(GNSS 1)* for aviation will be based on the US military's *GPS* (Global Positioning System) with other forms to follow as the world expands the commercial use of this technology in areas other than aviation.

Guidance - The introduction of digital avionics to the cockpits of military and civil transport aircraft occurred over twenty years ago, and has resulted in various attempts to automate the piloting activities in the "glass" cockpit of modern aircraft through what has become known as the *Flight Management System (FMS).* The current forms of *FMS* can be extended to provide an *ATM-compatible* version *(AFMS)* which will provide a new automated guidance capability which can be called *precision flight* along complex dynamically pre-determined, complex paths in the face of changing winds. The guidance capabilities of aircraft will be classified by a guidance performance rating (inaptly called Required Navigation Performance *(RNP))* which classifies their accuracy to conform to complex assigned paths. AFMS systems will become cheaper allowing their introduction of Glass Cockpits into smaller general aviation aircraft in the next thirty years.

Surveillance - The new technologies for surveillance are the datalink technologies since it is planned to introduce "extended" surveillance information dependent on the airborne measurement of aircraft position and other digital information about both the current state and the intentions of aircraft. Independent surveillance will be retained in the high density Terminal areas using MODE S secondary surveillance radars, (and probably backed up by primary radars needed for weather surveillance), but *Automatic Dependent Surveillance (ADS)* will become available in various forms (ADS-M, ADS-S, ADS-V, ADS-H) where M uses MDL, S uses SDL, etc.). By controlled request from the ground, the datalinks can automatically provide extended surveillance information from each aircraft at varying rates required by the operational situation.

Based on the squitter of the Mode S transponder there is a form of datalink called ADS-B (B = broadcast) which broadcasts high quality extended surveillance information between aircraft with Mode S transponders and to Mode S ground stations.. The installation of a Mode S transponder will give every aircraft a unique identity which is automatically transmitted along with the onboard measurements from GNSS of position and altitude. However, it will be possible using the digital avionics of the AFMS to extend surveillance information to provide some measure of current state such as path rates (ground track direction/groundspeed/vertical rate) as well as intended path information (next waypoint/next altitude/next direction).

The provision of this *extended surveillance information* is an extremely important new capability since it promises a means for the reliable and safe automation of high quality Separation Assurance, thereby creating the feasibility of many new concepts for tactical ATC operations. Remember that ATC functions and processes are dependent on intended information, **F**, not the information on the past path. Finally, with the advent of ADS-B, there is the promise of using intention information in ATC!

Automation - Although not recognized by the ICAO FANS Committee the final new technology is that of high quality color computer displays which when coupled with various forms of software automation called *Decision Support Systems (DSS)* will allow the introduction of Human-Centered Automation at the ATC controller's console. This is a key to increasing the productivity of ATC controllers and reducing the cost of operating ground-based ATC systems. Hopefully we have learned from our twenty years of trying to automate the cockpit, and will accomplish this in a safe and efficient manner.

We are now in a position to talk about some of the new operational concepts for ATC. They can comc in various forms - from small compatible changes in suboperations, to completely new operational environments for providing the basic ATC functionalities.

7. Examples of New Concepts for Air Traffic Operations

Example 1 - *Airborne Separation Assurance System- ASAS*

The provision of ADS-B information on the intended path of nearby aircraft will allow the design of a safe and reliable cockpit DSS for automating the airborne Hazard Management function between aircraft equipped with Mode S transponders and AFMS equipment. There is research and development work needed to explore how this would be accomplished, and where significant benefits from implementing such capabilities would be achieved. Note that ASAS (Airborne Separation Assurance Systems) are different from ACAS (Airborne Collision Avoidance Systems such as TCAS) in that they are not designed to act as a backup against errors in the ground based Separation Assurance function using last minute avoidance maneuvers. The design of a ASAS not a simple problem, even at the conceptual level of analysis, as perhaps the following example of an encounter will illustrate.

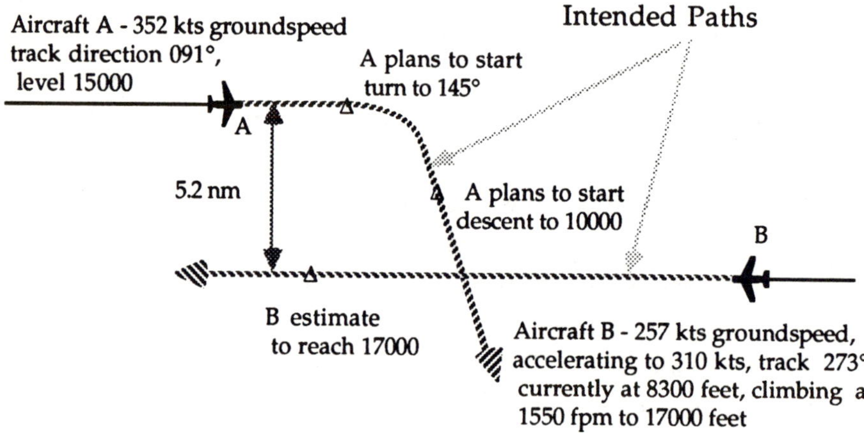

Figure 4. Example of Aircraft-Based Separation Assurance (ASAS)

Figure 4 shows two aircraft at roughly 50 nm separation approaching each other on opposite parallel tracks where it would appear from an extrapolation of their current tracks that there is no potential hazard since they will pass with an estimated 5.2 nm lateral separation. However, aircraft A is arriving at its destination airport and has <u>intentions</u> to turn to the right and descend which would put it on a path which crosses in about five minutes, and also intends to start a descent after the turn is finished. Aircraft B is at 8300 feet, climbing at 550 feet per minute and accelerating, and while it has 7700 feet vertical separation at present, it intends to climb to 17000 feet and will be co-altitude with A in about 5 minutes. If neither aircraft had any knowledge of the other's intentions they would have no idea at this point that a hazardous encounter is probable although aircraft A may see that it is not advisable to perform its intended maneuver.

If ASAS is on board both aircraft, the intentions of A and B are broadcast and the automatic Hazard Monitoring process of each ASAS can try to estimate the vertical and lateral separations at the planned passage. The airborne Separation Criteria for such an encounter are dependent on the ability of each airplane to conform to its intentions, and there may be no need to intervene if both ASAS monitoring processes indicate that these criteria will be satisfied. If not, then one (or both) aircraft must modify its intended path.

There are a number of operational issues in introducing such operations;

1. Who is in charge of Separation Assurance? Ground? Aircraft A? Aircraft B?
2. If the ground ATM service delegates Separation Assurance responsibility for this encounter, does it assign Resolution responsibility to one aircraft? Which one? Must the other aircraft accept its passive role in resolution? Should a ground automated resolution process recommend a type of resolution maneuver when it assigns responsibility for each passage? Is the delegation of airborne Separation Assurance limited to certain encounter situations? (e.g. level flight? only opposite direction encounters of short duration?)
3. What happens if the passive aircraft becomes alarmed at the resolution path, or its lack of initiation? Can it maneuver? Is there a standard resolution maneuver recommended for different encounter situations?

4. If the ground has given a general, long term delegation for Separation Assurance to both aircraft, and does not assign responsibility to one aircraft in each encounter, how do the pilots negotiate a resolution? When does the negotiation start? When does it finish? Is it conducted by radio or datalink? Is there an automated resolution transmitted to both aircraft? What information on intentions would an automated airborne resolution system need (e.g. where is aircraft A going since it plans to turn and start a descent? Perhaps it is starting its arrival descent to an airport - is it critical that it not delay the turn or the descent?)
5. How are the air to air Separation Criteria established? What are the equipage requirements? How accurately can the expected separation at passage be estimated as a function of equipage, planning horizon, and wind gustiness? What are the training requirements for pilots? How is the reliability of execution of proposed resolution procedures established?
6. What happens if Separation Criteria are violated? Are the pilots jointly responsible? Is the ground monitoring the passage? Will there be a historical record of individual pilot performance in resolving hazardous encounters?

Example 2 - *Digital Vectoring of Aircraft in Terminal Areas*

If a Mode S datalink is available in the Terminal Area, it can be used to send dynamically adaptive "digital" vectors to aircraft for maneuvering to achieve accurate spacing of aircraft on final approach to one or more runways and to ensure separations for arriving/departing aircraft. The arriving aircraft can be assigned a digital *DAR (Digital Arrival Routing)* during their descent which describes a complex set of 3-D legs right to their assigned runway and includes a *PMA (Precision Missed Approach)* if it becomes necessary. Departing aircraft will be assigned a digital *DDR (Digital Departure Routing)*. These paths would be pre-loaded data available in the AFMS (ATM-compatible Flight Management System) and could be activated

by controller messages to the aircraft via datalink during arrival descent and taxi-out. But due to the unexpected exigencies of traffic handling for arrival and departing aircraft, the fourth dimension, time, cannot be pre-determined. These 3-D paths will be "fine tuned" by digital modifications which introduce timing and consequently the location of the turnpoints, speed changes, and climb/descent points. All digital messages from the ground might be generated by automated Metering and Spacing systems like CTAS (Center-Tower Automation System) and would be approved by the ground controller before being sent via the datalink. (see figure 5.)
Similarly, upon receipt in the AFMS they would be displayed in the cockpit for pilot approval and acknowledgment back to the ground, and then for entry into the aircraft's AFMS guidance process by the pilot as the aircraft's commander.

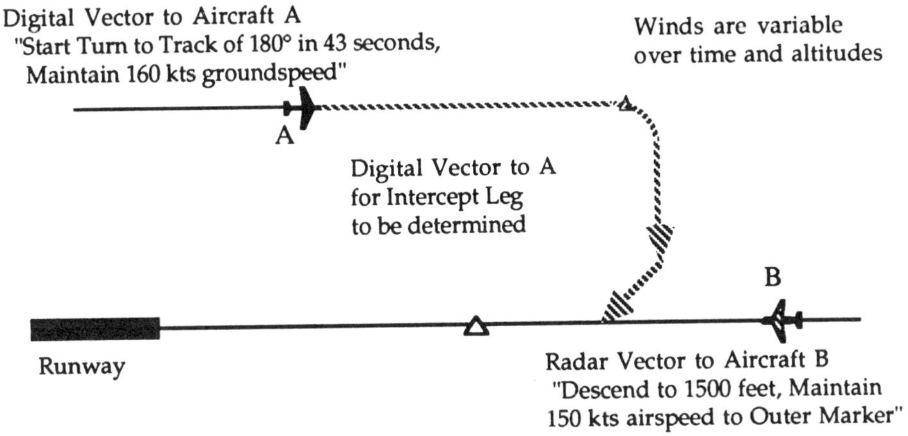

Figure 5. Digital Vectoring of Aircraft

But unlike Radar Vectoring where the controller radios up headings, airspeeds, and altitudes, the digital vectors would be much more precise specifying *ground track*, *groundspeed*, and *altitude rates*, and the *times for execution* to the nearest second. This eliminates the effect of uncertainties in the time-varying winds over space and altitude in the Terminal area and will lead to more accurate conformance to the times scheduled for runway operations by future automated systems which will be tactically adjusting to the changing traffic situation.

Digital vectoring brings new operational issues in specifying its implementation;

1. Because specifying the groundspeed and groundtrack means there will be an associated airspeed, the ground will need to have some approximate information about the winds and performance of each aircraft type, and realtime information about its current gear and flap configuration; e.g., the limiting airspeeds for the aircraft given its flap and landing gear configuration, or the maximum rate of climb given its weight, air temperature, and airspeed. This data can be available for each type of aicraft in the ground computer, but now configuration and weight data must be datalinked down.
2. Specifying altitude rates means that knowledge of each aircraft's current gross weight, and type of engine must be known. For maximum climb rates, the air temperature should also be known to some approximation.
3. What are the relative roles for pilot and controller when a digital vector can be sent directly from the ground to the AFMS? Can we see a mode of operation where the ground controller is directly guiding the aircraft with the pilot only monitoring the guidance commands? (Note this situation has existed for many years in Ocean shipping when the captain of the ship surrenders this responsibility to the local Harbor Pilot for entry and departure from harbors).

4. Can the Separation and Obstruction Criteria in the Terminal Area be reduced when aircraft are digitally vectored? (Note that many limitations in the design of terminal area airspace and procedures arise from the use of the ILS (Instrument Landing System) and its lack of precision guidance if a Missed Approach occurs requiring radar vectoring and changes in planned paths for departing aircraft).

Example 3 - *Human-Centered Automation for Final Approach Spacing*

As an example of a concept for automating the Scheduling process of Congestion Management, a brief description of the ASLOTS research project (see [2]) currently under way at the Flight Transportation Laboratory will be provided here. Its goal is to create an interactive, adaptive, automated Decision Support System for an ATC controller/planner to manage the approach and departure flows of traffic to and from a set of parallel or crossing runways. As shown in Figure 6 for a single runway only, ASLOTS shows the schedule plan for arrivals and departures as a graphic set of Slot Markers on a display before a controller who is planning the assignment, sequencing, and scheduling of takeoffs and landings to the active runways. This would be a new position in the terminal area facility. By manipulating the Slot Markers, the assignments, sequences, and timing of all runway operations within the next 15 minutes can be changed within feasible bounds which are shown to the controller/planner. It assists the controller/planner to instantly construct and modify the planned schedule of runway operations in an efficient manner; for example, the controller/planner can try inserting takeoffs between landings and the logic will open efficient gaps in the landing stream to accommodate the desired number of takeoffs.

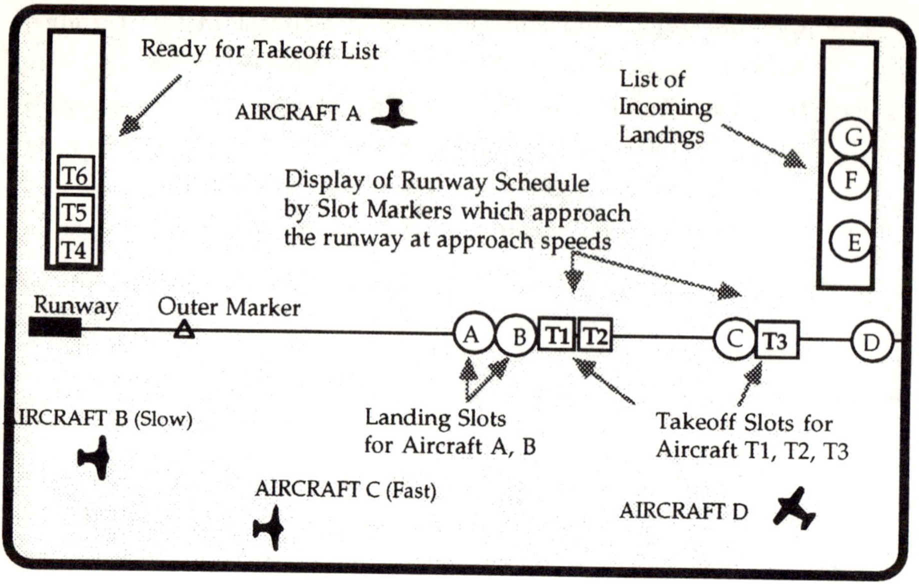

Figure 6. Runway Operation Planner's Display - ASLOTS

ASLOTS uses familiar patterns for arrival and departure as discussed above to dynamically construct and re-construct detailed, conflict-free, flexible 4-D intended paths for all arrivals and departures. The flexibility of the patterns allows the paths to be modified within limited ranges and arrival or takeoff times to be adaptive to aircraft conformance errors due to wind or piloting, or to unexpected operational events like missed approaches or lengthy runway occupancies after landing which prevent a planned takeoff. As the paths are adapted, a set of digital vectors are generated which can be shown graphically to the final spacing controller (another person, not the runway planner), who is responsible for sending them directly to the aircraft via radio or datalink at appropriate times. The planner's position shows all runways while each final approach spacing controller has a reduced display. The goals of such automated systems are to increase the effective use of runway capacity available in bad weather while also reducing the Final Spacing controller's workload.

Example 4 - *Dynamic Flow Control*

To illustrate a new concept for Congestion Management, another research project at the Flight Transportation Laboratory called *Dynamic Flow Control* (see [3],[4],) will be briefly described. It is concerned with integrating several forms of traffic flow management controls over the arrival flow at a congested airport over a period of reduced capacity due to bad weather. It assumes that there is realtime surveillance information about the progress of all airborne aircraft inbound to the airport, knowledge of all flight plans for aircraft still on the ground, and knowledge of schedules for aircraft which are likely to file flight plans at the airport during the day but which have not filed a flight plan yet. It also assumes that there will be a good means of instantly communicating with all inbound aircraft and/or Airline Operations Control Centers (AOCCs).

The Traffic Flow Controls used are *Ground Holds (GH), Air Holds (AH),* and Enroute Controls such as *Cruise Speed Advisories (CSA), Controlled Time of Arrival at Entry Fixes (CTAF), Controlled Altitude Advisories (CAA), and Miles-in-Trail (M*-I-*T).* These are used in an integrated manner to dynamically control the arrival flows based on 15 minute updates of a Traffic Flow Plan based on the actual conditions seen at the airport, the actual holding at the air holding points, and the actual position of all enroute aircraft. Since it is difficult to forecast the weather conditions and the time varying Airport Acceptance Rate, (AAR) and since new unexpected unscheduled flight plans may arrive while others are being cancelled, or the airport may suddenly only be open to ILS Cat. 2 or 3 aircraft, or since aircraft may be early or delayed in getting airborne compared to their stated flight plan times, there is a need to keep revising the Traffic Flow Plan and to re-estimate the probable air holds at/around each Entry fix, and issue revised Ground Holds and enroute cruise speeds.

To allow interaction by the Traffic Flow Managers, there are controls which limit the flow rates at any fix; to eliminate, limit, or minimize the air holding at any Entry fix; and to limit the number of Traffic Flow Advisories issued at any 15 minute update. It will not issue small changes in desired enroute speeds or small changes in desired CTAFs, nor will it issue Ground Hold changes smaller than 15 minutes. A 30 minute warning on Ground Holds is guaranteed so that if an aircraft passes within 29 minutes of its controlled departure time, it is able to commit to a takeoff, become airborne and then be subject to the enroute controls. Enroute, the aircraft can have its airspeed controlled within pre-determined limits as required to maintain an efficient arrival flow, or can be asked to make a CTAF if it is feasible. It may be asked to fly faster whenever a gap appears in the planning of arrivals due to better AAR, etc. In this way delays can be reduced since the airport is not allowed to become idle when it should be busy.

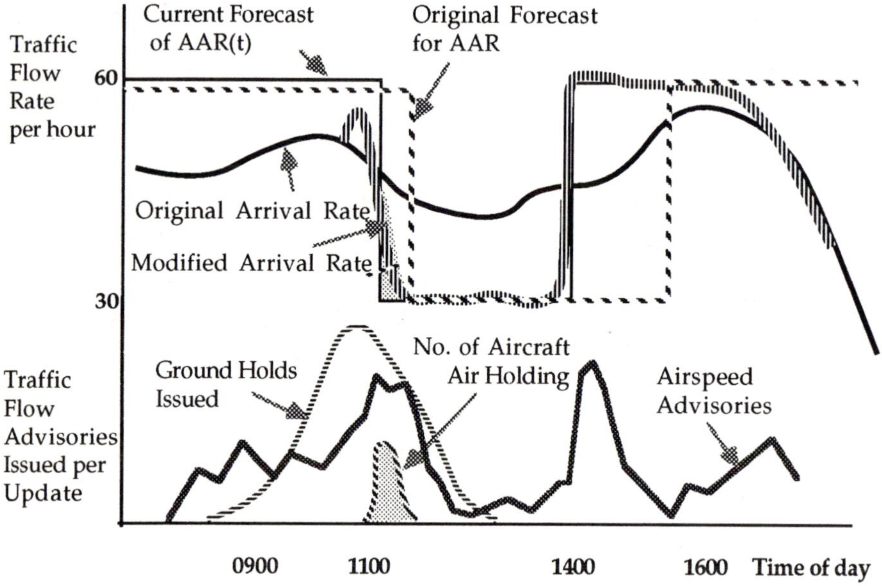

Figure 7. Dynamic Flow Control

Figure 7 shows the dynamic nature of the issuance of Traffic Flow Advisories and the typical behavior of air holds and ground holds for a scenario where there is a significant reduction in Airport

Acceptance Rate for a period of three hours. Note that DFC plans both the reduction and the increase in the airport's actual arrival flow rate to match the expected AAR, it can increase the arrival rate in the period just before the AAR is reduced using increased airspeeds, and is busy working off the backlog in the period after the AAR returns to higher values.

8. Conclusions

A framework has been created to describe ATC functions and processes, and the flow of information between those processes. It can be used to describe any concept of ATC operations, and is necessary for designing automation and human activities in that concept.

The task of engineering a new concept for ATC operations using the available proven technologies is not simple, and requires considerable analysis and simulation to validate the design and to demonstrate it to the aviation community. There are many feasible new concepts using the current wealth of proven technologies.

The new ATC system based on ICAO FANS technologies is global in its scope, and will introduce automated Decision Support Systems for the ATC controller. While these can be implemented technically in slightly different ways, the ATC procedures used around the world must be identical for international pilots and there must be a common equipage for aircraft.

There will be a long transition period for introducing the equipage and procedures. It is vital that new procedures be able to safely operate simultaneously with current procedures.

References

[1] Robert W. Simpson, *Engineering of Air Traffic Control Systems*, Vols. 1,2 unpublished notes for MIT course 16.72.

[2] H. Idris, *Human Centered Automation of Air Traffic Control Operations in the Terminal Area*, FTL Report R94-8, Flight Transportation Laboratory, MIT, 1994.

[3] F. Fedida, *A Dynamic Approach for Air Traffic Flow Management of Arriving Aircraft at a Congested Airport*, FTL Report R 94-2, Flight Transporation Laboratory, MIT, 1994.

[4] G. Wong, *Evaluation of Dynamic flow Control over Aircraft Arriving at a Congested Airport*, SM Thesis, MIT, February, 1996.

THE PROBABLE EVOLUTION OF THE "AIR SEGMENT" ON A MEDIUM AND A LONG TERM BASIS

Marc Pélegrin

Member ANAE and FEDESPACE
1, Avenue Camille Flammarion
31500 Toulouse, France

The expected traffic evolution is sketched and some consequences are considered. It is suggested to include in the "Air Segment" the movements of the planes in the airport apron. The automation of the movements seems a necessity. Three major factors interacting with the flights are considered: the possibility of a fully automatic flight, the use of an unique worldwide navigation and landing system and the concept of "free flight" (or autonomous flight). Finally, the possibility of a global optimization in a large zone is discussed.

1. General Assessments and Definition of the "Air Segment"

1.1 Worldwide traffic expected growth

In 1990 a study commissioned by EEC/DG XII [1] accepted a worldwide growth rate of 2.7 % to 2.9 % for economic activity. Air transport demand was estimated to rise by 4.5 % per year - implying an almost threefold increase in air transport demand in the next 25 years - while cargo traffic was expected to grow by about 8 % per year.

In a EUROCONTROL report entitled "Time-based Air Traffic Control in an Extended Terminal Area" [2] a summary of the expected rate of growth as perceived by the Organization was given: average annual passengers growth evaluated at 6 % in Europe with a peak at 10 % around 1999. At such a rate the number of passengers would double by the year 2000 and triple by 2010.

The rate of growth of freight was quite linear between 1985 and 1990 (about 12 %) with a temporary stagnation around 1991.

Taken into account that some aircraft for domestic traffic or short distance traffic - let's say between 700 and 2000 km - are becoming larger and larger, Boeing and Airbus, in 1994, agreed that the accepted rates of growth will lead to doubling the aircraft movements around year 2008 $\pm$ 2 and tripling around 2015-2020.

After 2010, the market will be a replacement rather then a supplier's market: the rate of growth will certainly drop by 0.5 % or 1 % and later by 2 to 3 %. However, the Asian market has a rate of growth much more important than the Europe/US ones and it may compensate for the slowdown mentioned above.

Another factor will operate in the same direction: the "guaranteed" aircraft life, which was 30,000 flight hours until recently, has now increased to 35,000 hours, and 40,000 hours are expected in the case of aircraft built in the years 2000-2010.

Short-haul traffic is expected to decrease *at least in Europe* and in some other congested parts of the world owing to the rapid development of rail and highway networks. Scheduled air traffic between cities 400-600 km apart will gradually disappear over the next 15 years.

Moreover, the rapid introduction of high-speed trains (TGV) suggests that air services between cities 400-700 km or more apart will no longer be available either. However, acceptable transversal accelerations for high-speed trains limit the radius of curvature of the track on both the horizontal and vertical planes; hence, where there are mountains between cities, the cost of the track may be too high and TGV links may be abandoned.

1.2 Air transport constraints

The major factors affecting air transport development are:

a) noise restrictions; they will have repercussions on the development of new aircraft and their use. Noise pollution in the vicinity of airports is defined by ICAO; sometimes, restrictions around airports are more severe than the ICAO regulations.

b) emissions (exhaust) from aircraft engines are potentially harmful additives to the atmosphere. In Sweden, the landing fees are adjusted according to the type of engines the planes are equipped with. Aviation consumes 13 % of the total fuel consumed in transport worldwide; the fact that a subtantial amount of this fuel is burnt at the tropopause level or higher may create problems (CO_2 NO_x and No_y). In consequence, new restrictions may be put on ATC in order, for example, to limit the number of aircraft per day at each flight level (above FL 350). Supersonic aircraft would be fully concerned because they fly above FL350.

c) airport extensions (additional runways) are very limited in Europe (environmental constraints). In many large cities the only possibility is to build a new airport from scratch at a considerable distance from populated areas (this is the case of the new Munich airport).

The main task for the near future is to make optimum use of airport capacity (landings and takeoffs). Two improvements would be:

-optimum delivery of aircraft to the entry gates to coincide with takeoff;

-optimum time slot allocation combined with reasonable distribution of flights (agreements between airlines).

Only an overall approach will result in airports capable of coping with the traffic expected within the next 15 years.

d) coming back to the second-generation supersonic aircraft, it is clear that their integration into ATC will (...or would) imply precise algorithms in order to avoid any "holding patterns". At the present time, the only supersonic civilian aircraft are the 13 Concorde, of which no more than two or three fly concurrently; their routes and schedules are different, their time slots known 12 hours in advance, wind interaction is smaller than in the case of subsonic aircraft, thus making them easy to manage.

If the second generation finally arrives, there will be some 300-500 aircraft to monitor (it means that, at a given time 60 to 100 aircraft may be flying, some of them on the same routes and on tight schedules. If the ascent cruise, accepted today in the case of Concorde, is an optimum solution, it will entail precise aircraft guidance. Fortunately, new technology, such as the Automatic Data Reporting System (ADRS) will soon be operational (in any case, before the second supersonic generation appears).

As it is inconceivable that priority be given to supersonic aircraft, with regard to subsonic ones the relative importance of delays around airports penalises supersonic aircraft much more than subsonic aircraft on an airport-to-airport flight. "Linear holding", which consists in requesting an aircraft to fly from supersonic cruise to subsonic cruise at a given time or position, seems to be the only possibility of managing a mixed fleet of subsonic and supersonic aircraft. However, if we move towards Air Traffic Management, it seems that only global penalization would solve the problem. By global penalization we mean a criteria which takes into account the sum of all the penalizations attached to each aircraft flying inside the considered airspace. The optimization criteria consists in minimizing the sum of penalizations over a given time horison. Agreements should be signed between airlines which operate supersonic aircraft and those which do not.

e) large capacity aircraft (600-800 passengers) are being studied by the two largest manufacturers in the world, Boeing and Airbus; it is worthwhile noting that both partners, in June 1995, have claimed that the project should be delayed because of the market evaluation. The size has been determined by basic airport requirements, namely runway, taxiway, and platform configurations: the gate systems - if any - would require modification. For example, if an aircraft has two decks, each deck should have its own gate, thus entailing a total of 4 or 6 gates per aircraft.

Many other problems must be solved in order to handle so many passengers in about the same amount of time as is required today to handle 400 passengers arriving on one aircraft. The advent of such large aircraft will again raise the problem of priorities: what will the rule be when saturation is reached ? Do we ignore aircraft capacity and organise traffic in such a way to achieve minimun total waiting time for the *aircraft* concerned, or do we minimize total waiting time in respect to all *passengers* in the sky ? This is an ongoing problem, which will be increased by the advent of a large capacity aircraft.

f) a word about cargo: its growth rate of about 8 % per year is being sustained despite the current economic crisis. In tonne-km, air-mail represents about 9 % of the total. However, it seems that the growth rate for mail is lower than for goods, and that this trend will be maintained in the future. The "quick-change" aircraft, i.e. aircraft which carry passengers during the day (domestic traffic) and mail at night is an interesting experiment; the "change", i.e. the removal or the rearrangement of seats, is carried out on small carriers (50-100 seats) within 45 to 60 mn.

In the future, it will be necessary either to monitor both the cargo and passenger areas at airports or to specialize airports to cargo. They both give rise to similar problems; docking (the equivalent of gates), services, connecting containers or goods, connection to railways. However, some problems such as temporary storage (sometimes cold storage) are specific to cargo transport.

1.3 Air transport optimization

For the benefit of all, the traffic management should be optimized in a way that a "generalized cost" should be minimized. This is not a straightforward concept !

First, let us consider a plane which will be alone in the sky. The optimal flight is obvious, it results from a compromise between the time of flight and the fuel consumption.

In the real case, such an optimal profile could not be attained, there are constraints on the profile of all the planes in a given airspace; then, it is necessary to look for the minimum penalization cost for all the planes which will use this airspace. But traffic is a continuous process and the time horizon should be fixed in order that optimization algorithms could be applied. In a very simplified form we can say: how to place the planes in the given airspace (FL, Mach, rate of climb, top of descent, descent profile...) in order that the total penalization due to the deviations from the optimum flight profile (plane alone in the sky) be a minimum over a specified time horizon, half an hour for example.

At the present time such a problem has no solution on a real time basis. However, we can assume that it will have one within the next 10 or 15 years. (see parag. 4)

The optimization process implies three major topics:

a) to define the boundaries in which the system to be optimized stands
b) to elaborate a "generalized cost" or "penalization function"
c) to find the "best arrangement" which minimizes the global penalization function over a given time horizon.

1.4 "Air Segment" definition

From the comments given above, it follows that movement of the plane on the airport platform should be included in the "Air Segment" because more and more the ground traffic will be a blocking factor with regard to the landing and takeoff rates. In others words, *we defined the "air segment" as the control of the plane from "gate to gate"*.

Before concluding this paragraph, let us say a word about air safety. Air safety is good. However, progress should be maintained for at least, two reasons:

1) even if no major innovations occur on the aircraft concept, the doubling of the number of civilian planes is expected around 2008-2010 and the tripling around 2015. At the present time, one mid-air collision occurs every 3 years (Russia and China not accounted for). If the number of planes triples it is quite obvious that the number of mid-air collisions will be multiplied by a factor greater than 3.

Assuming that the *safety factor is kept at the same current value* (better than 10^{-7} per hour of flight), more than one mid-air collision will occur every year, a positively unacceptable situation... Then the safety factor must be improved in order to keep the mid-air collision risk at the present level or, better, at a lower level. The major cause of mid-air collisions is due to errors by Air Traffic Control; however, the interaction between the atmosphere and the aircraft may contribute to these errors.

2) very large aircraft (600-800 passengers) are considered today. First investigations show that the frequencies of the wings and fuselage modes will be of the same order of magnitude as the ones resulting from the control laws (pitch, roll and yaw) of the aircraft. New developments concerning the actuators and maybe the mobile surface principle to generate forces will perhaps be necessary. It is clear that the interaction between the atmosphere and the plane will be quite different from what it is nowadays, even on the largest aircraft (the horizontal rear empennage of the 600-800 passengers plane will be approximately the wing of a present A 320 !) - The symmetrical "certification gust" will no longer be acceptable, unsymmetrical gusts applied to the wings should be considered; special attention will concern the wing vortices, mainly during landing and takeoff. The structure deformations in any phase of the flight, and especially during landing or takeoff under severe microbursts should be entered in the control-computer and compensated for.

2. Automation of the Ground Movement Phases

2.1 The obsolescence of the present airport organization

It is thought that movements on the airport taxiways and runways are already a limiting factor for the takeoffs and landings in certain airports, namely in Europe. This occurs mainly in bad weather conditions. Most of the major European airports are equipped with ILS CAT III landing aids, a system which allows automatic landings even if vertical visibility is reduced to zero.

Planes should be equipped with redundant equipment to comply with safety, even after a first on-board failure. To comply with the landing rates attained in bad visibility, it is necessary to improve the ground movements of the planes, namely in bad visibility to and from the gate.

Gates availability is already a limiting factor. To extend the size of the buildings of the terminal is no longer a solution. Gate occupancy should be limited to the passengers' embarkment and disembarkment. Aircraft servicing (cleaning, refueling and catering) should be provided outside the gate: a second reason for automatic movement on the ground.

Moving an aircraft on the ground with the jet engines becomes an abnormal situation when efforts are made to limit the consumption of fuel. A medium size aircraft, (160-200 passengers) consumes 360 kg/hour per engine when taxiing. If 30 mn elapse between the start up and the arrival at the runway threshold, more than 300 kg of fuel have been burnt for a distance of 2 or 3 km ! A tractor with a diesel engine would use less than 20 kg of fuel. This third reason should be kept in mind by the manufacturers and airport managers at a time when ecological pressure is increasing rapidly.

Airports are now surrounded by technical zones and even houses. Noise level is defined by regulations. The present one is FAR 36, Chapter 3. Noise around an airport should not exceed a given value expressed in EPN dB (Equivalent Perceived Noise). The regulation concerns only maximum levels of noise. It is probable that, in the near future, continuous low level noises will be taken into account. The noise produced by planes moving on the runways or taxiways will then be considered.

Aging of components, namely jet engines and brakes, during ground movement is not negligible. Carbon disc brakes are more degraded when they are cold than when they are hot. When the plane leaves the gate for the takeoff runway, brakes are cold; if planes are queuing then the brakes are constantly used. This should be corrected.

Automatic control of all the movements on the ground will allow all kinds of interchange while queuing; an optimization of the tractors movements, thanks to a central computer, even in bad weather situations, will be achieved.

2.2 Automatic tractors

A manually operated system could be an intermediate step. However, the management of a fleet of man-operated tractors will be difficult; safety should be guaranteed at the same level as today.

Let us consider some of these difficulties:

a) to be efficient the plane should be towed immediately after it has exited the runway.

For manual operated tractors, it is difficult to envisage a rendez-vous without stopping the plane (this will not be the case with an automatic system). It is estimated that from the beginning of the braking phase to the time at which the plane is moved by the tractor at 10 kts (the nominal speed) some 45 to 60 s have elapsed. Such a delay may interfere with the landing rate which sometimes corresponds to 50 - 60 s spacing.

b) the non-uniform distribution of the airport traffic during a 24 hour period will lead to an important number of operators with peaks and unloaded periods.

c) the operators should always be instructed by a central office and, in case of fog, should follow the instructions without any failure in spite of a very low visibility. What about fatigue ? Tractors should be automatically located and instructed. The presence of man does not seem necessary (Note: such a system will not « produce » unemployed people; at the present time, it does not exist on a manual basis. This would be a new system. Jobs will be created to conceive, realize and operate the system). Then it is suggested to develop a fully automatic system. The system proposed here has been detailed in [3].

The objectives are:

- to provide an automatic rendez-vous between the tractor and the plane without stopping it (nominal speed 5 kts).

- to "catch" the plane by the front landing gear complying with the force limitation issued by the plane manufacturer.

- to optimize movements on the airport according to a criterion which tends to maximize the number of movements per hour.

The criterion should be modified at any time to compensate for an incident or a justified priority.

- to guarantee, at least, the same level of safety as today ($0{,}5.10^{-7}$/h of operation with passengers on board).

Due to the fact that the front landing gear is a "weak" part of the plane, it is proposed to make the rendez-vous in two steps. The tractor is, in fact, composed of two vehicles called A and B. Vehicle A is a heavy one (mass of the order of 25 T); it has the capacity of towing the plane; vehicle B is a light one (mass less than 1 T), its power is just sufficient to provide its mobility when alone; it has some intelligence on board.

The sequence is (Fig. 1):

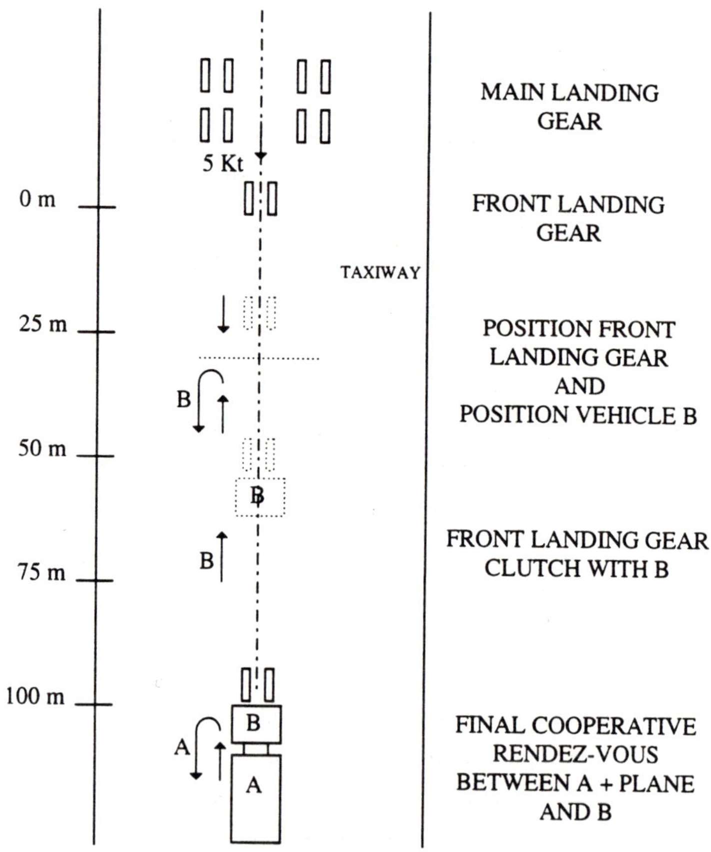

Figure 1. Phases

The plane which has just landed is supposed to exit the runway through a specific taxiway; on this taxiway sensors check that the plane passes over them and check the speed. If these data are correct, then the tractor i.e. vehicle A + vehicle B attached, goes on the taxiway and stands there, right on the center line. Vehicle A has a precise, absolute, navigation system. The vehicle A liberates vehicle B. This one navigates towards the plane, which is arriving using sensors located on A and B. At a given distance of the plane - a distance known by the control room - B gradually shifts from a navigation relative to A to an intelligent navigation based on the reconnaissance of the front landing gear of the plane. In any case, when the distance between vehicle B and the front landing gear reaches a given value the movement of B is inverted; B is now going in the same direction as the plane, but slower.

Now let us suppose that B and the plane are attached. B becomes immediatly passive and the pilot still controls the plane (and consequently B).

The last phase of the rendez-vous is now engaged. Vehicle A will navigate towards B: this is a *cooperative rendez-vous* because A and B know each other perfectly. In other words, the role of B is to fix up sensors on the incoming plane which can be analyzed by the sensors fitted on A.

Automatic control of the planes and, by extension, of all the vehicles which run on the apron will lead to minimum transit time; in addition the gates will be used only during the time requested by the passengers displacement. Planes could be pushed to a specialized area for servicing (fuel, cleaning, catering...).

The localization of all vehicles on the apron will result from GPS; the D-GPS and if necessary the "phase pursuit" on one of the carrier frequencies (L_1 or L_2) may be used without difficulties because "reference points" are easy to specify on the airport platform.

3. Automation of the Airborne Phase

3.1 The atmosphere

The ATC works on ground velocities to elaborate a strategic arrangement of planes in a given airspace. The vertical separation is arbitrarily referenced to a barometric pressure (1013.25 hPa). The horizontal separations should be reduced in the future to cope with higher densities of planes in a given airspace. The horizontal gradient of pressures, winds and temperatures should be taken into consideration in order to guarantee a minimum separation distance in the isobaric surface on which the plane evolves; a compromise should be derived between the real ground velocity of the plane (which will be known on a real time basis soon) and the necessity to avoid air velocity changes too frequently. On board, on most aircraft, the IAS (Indicated Airspeed) or the CAS (Calibrated Airspeed) are presented on the instrument panel; on modern aircraft other parameters are computed as we will see below.

The real atmosphere departs from the standard one. The atmosphere is composed of the *troposphere* - a 8,000-11,000 m high layer around the earth - in which there are vertical movements namely in active cumulo-nimbus clouds and the *stratosphere*, just above the troposphere, in which air movements are mainly horizontal. The separation, called *tropopause*, is a transition layer in which both characteristics are present. The position and thickness of this layer varies with the latitude and the seasons.

In the troposphere, *strong turbulence* even in clear sky may exist. Active cumulo-nimbus are dangerous for a plane due to turbulence, vertical velocities (20 m/s, or more, in the core of the ascending mass of air/water), lightnings, icing...

In the stratosphere, *jet streams* are frequent; there are "tubes" of air of some hundreds of meters or of a few kilometers in diameter; the flow can be laminar and clear but the transition zone is highly turbulent.

The only measurable parameters on board tied to the atmosphere were static pressure p_s, dynamic pressure p_d and total temperature T_a. From these data *true airspeed (TAS)* and Mach number should be derived by St Venant or Rayleigh formulae according to the Mach number. The local wind around the aircraft is derived from the *true airspeed* and the *ground speed* if this one is available on board (Inertial Systems and/or precise localisation systems). The correct equations to be solved are:

- *subsonic uncompressible flow*: assuming that the local static temperature is available, the *true airspeed* is defined by:

$$V^2_{TAS} = \frac{2a^2}{\gamma - 1}[(\frac{p_d - p_s}{p_s} + 1)^{\frac{\gamma-1}{\gamma}} - 1]$$

where a is the sound velocity corresponding to the local static temperature and γ the ratio between the two heat coefficients.

If the static temperature is not available, the *calibrated airspeed* is defined by:

$$V^2_{CAS} = \frac{2a^2_0}{\gamma - 1}[(\frac{p_d - p_s}{p_0})^{\frac{\gamma-1}{\gamma}} - 1]$$

where a_o and p_o are the sound velocity and the static pressure at sea-level for the standard atmosphere.

- *compressible flow*: the Mach number is defined by:

$$\frac{p_d}{p_s} = [\frac{(\gamma + 1)^2 M^2}{4\gamma M^2 - 2(\gamma - 1)}]^{\frac{\gamma}{\gamma-1}} \frac{1 - \gamma + 2\gamma M^2}{\gamma + 1}$$

It is important to know the TAS or, at least the CAS and the Mach number, in case of flying in a windshear zone, in order to avoid getting out of the flight envelope of the plane. Nowadays the TAS is computed or extracted from stored tables and presented on the instrument panel.

In the near future, data, measured on board, will be automatically transmitted to the ATCs; the ground speed (which is now available on board thanks to INS or to precise localisation systems such as GPS) will be known accurately by the controllers and their electronic aids.

In the medium future (10 years), we can assume that GPS (and maybe GLONASS) will be the unique worldwide navigation system. In dense areas (Europe, United States , Asia), the WAAS (Wide Area Augmentation System) will be in operation before 10 or 15 years. (The USA will be covered by such a system in 1997-98 - contract Wilcox-Thomson); the localisation will be known within 10 to 20 m (in X and Y) according to the geographic profile of the overflown site.

Over the ocean or unpopulated areas, the GPS with *integrity control* thanks to communication satellites (initially INMARSAT) will give a precision of the order of 100-200 m (X and Y). The CE-GPS (Complement European - GPS) is under development.

Landing with D-GPS is already achieved for CAT 1; many developments are performed to try to reach CAT 2 and if possible CAT 3. No final date can be specified now for CAT 3, - the integrity problem should be solved - though some in-flight experiments based on a "phase following process" (OMN/Marker) have been conducted successfully at Standford University.

Before 10 years, every plane will be localized precisely on a quasi-permanent way (data refreshed every 10 secondes); air traffic control will be able to reduce the present separation distances and the data which will be fed into the on board computers will be homogeneous all over the world.

3.2 Automatic management of the aircraft

3.2.1 Present state

All the data involved in the flight mechanics are - or could be - available on board; in any case the flight coordinates in the flight envelope (Fig. 2) must be inside the V_{PROT} - V_{MMO} (or M_{MMO}) domain.

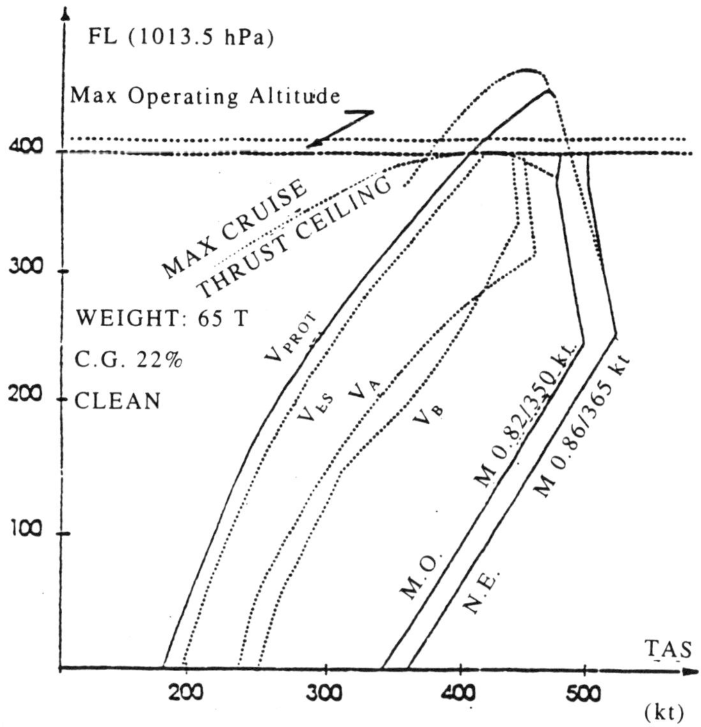

Speeds definition

L_s: lowest selectable speed. It corresponds to 1.13 V_s during take-off or following touch and go. It becomes 1.23 V_s as soon as any flaps/slat selection is made.

V_a: design speed of maneuver,

V_b: design speed for maximum gust intensity and rough air speed.

High speed protection

MMO/VMO: M. 82/350 kt,

MMO + 0.04/VMO + 15kt: maximum steady speed with full nose down stick.

High angle of attack protection

V_{prot} minimum speed (corresponding angle of attack: alpha prot),

If alpha prot is exceeded, the angle of the attack returns to and maintains alpha prot. This protection has priority over all other protections.

Figure 2. Flight Envelope (A320)

In the Airbus family, these limits are automatically controlled and override the pilot actions if necessary; in case of severe atmospheric conditions, the pilot receives a warning and, if necessary an on-board computer controls the throttle. Let us take an example:

According to FAA-AC25-12, on-board warning should be provided. The Airbus philosophy first defines a wind severity factor SF which reflects the instantaneous loss of energy due to the global shear (longitudinal and vertical). [6].

$$SF = \frac{dW_x}{dt} - \frac{g}{V} W_z$$

where:

W_x is the longitudinal wind (negative if headwind)

W_Z is the vertical wind (negative if downwind)

V is the indicated airspeed of the plane.

SF is filtered and compared to a fixed threshold, typically 0,13 g (which corresponds to 2,5 kts/s). A 4 s lag was found the best compromise to avoid false alarms with a probability less than 10^{-6} per landing. This value corresponds to the probability of encountering a windshear (at least in the USA).

However, for better warning and an escape manoeuvre capability if necessary, the warning system is a compromise between SF, actual aircraft energy and a safe minimum energy, this is the *wind shear warning* (WSW).

Let us consider the A 320 (Fly by wire). The pitch control law is C* up to a specified angle of attack normally the one which corresponds to 1.13 V_s (V_{stall}) (at 1 g); above the control law is an "angle of attack (AOA) control law" (Fig. 3) which is adjusted in a way that the maximum acceptable AOA prior to stall, corresponds to the full deviation of the side stick. In this configuration the AOA protection threshold will engage full power automatically. The angle of attack is estimated as indicated on (Fig. 4).

If a windshear occurs, the pitch control law depends on the requested load factor and tries to achieve a value less than 1 g (Fig. 5).

If the windshear warning is activated, the required control is "pull up to the maximum deflection of the stick and maintain it". It is worthwhile noting that even if the aircraft flight path has a negative slope, the above procedure is the best.

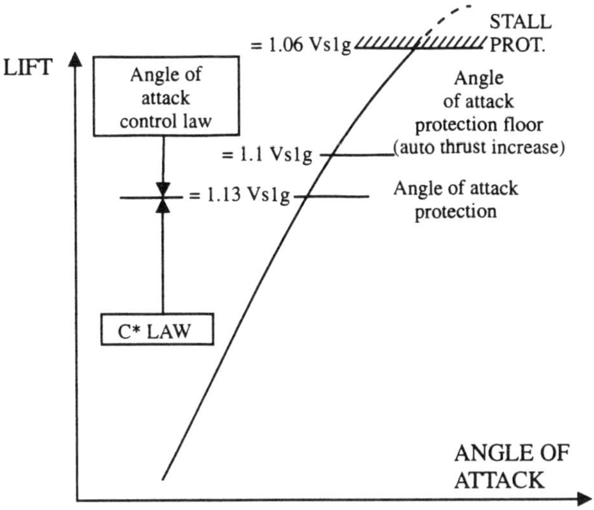

Figure 3. Airbus Control Law
AOA-protection at Low Speed

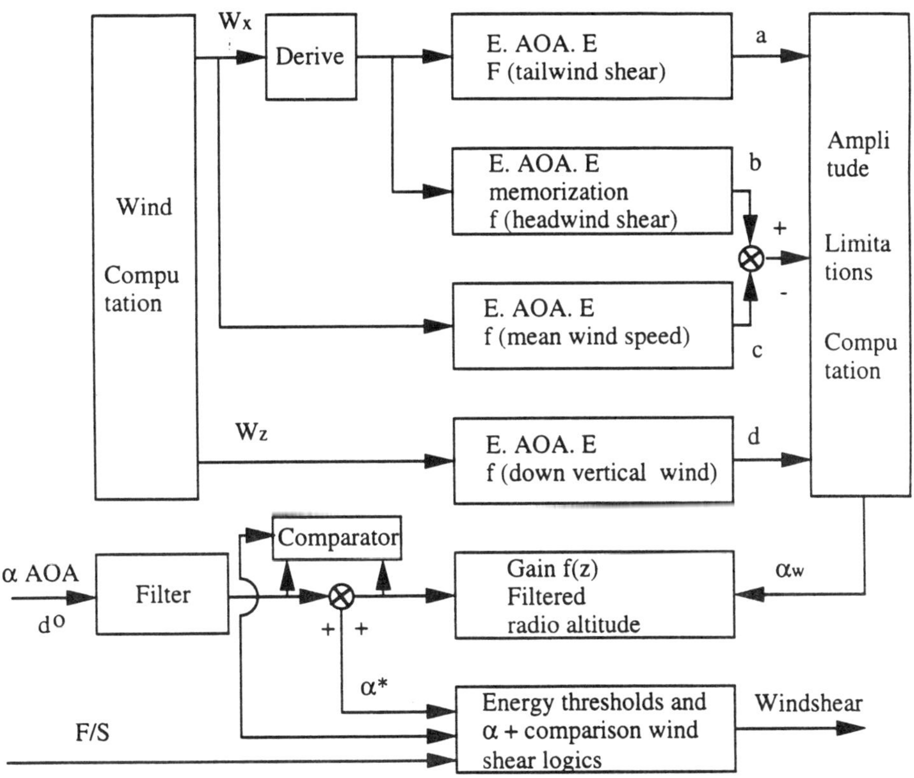

Figure 4. A320 Control Law:
Windshear Warning Computation Principle

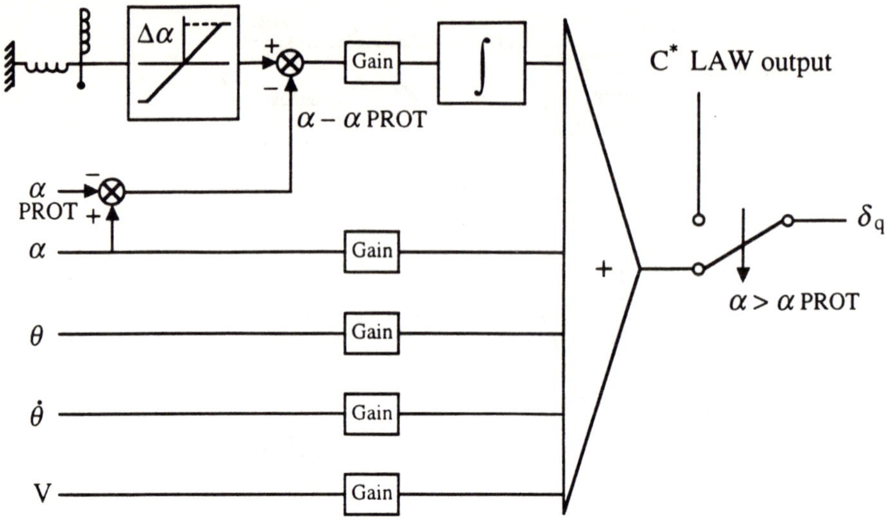

Figure 5. Control Law in a Windshear

Many simulations have been made after this accident: it has been proved that if full power had been maintained after the maximum height (when the plane was already descending) and an attitude control such that the AOA was 2 or 3 degrees below the stall AOA, the aircraft would have passed the microburst with a minimum height of a few meters (this is not a criticism of the pilot because at the time, microbursts were not identified as they are today).

Then, it is clear that the detection of a "high probability that the plane is flying towards, or in, a microburst" is vital. The knowledge of the ground speed, which is available on modern aircraft equipped with INS, facilitates the detection of the wind speed fluctuations. The larger the fluctuations in amplitude and frequency, the higher the probability of entering into a winshear.

At the present time, it does not exist reliable sensors to detect such atmopheric phenomena, neither to detect wind vortices due to another aircraft; meteo radars are quite inefficient, but for active clouds (cumulo-nimbus) because there are droplets.

In conclusion of this paragraph, it is clear that modern aircraft are well protected against meteorological perturbations; however for safety requirements the instructions given by the controller could be disregarded temporarily by the pilot.

In the next future, when the automatic data-link will be in operation, the instantaneous "state vector" of the plane will be known by the ATC and, in addition, the local wind and static temperature. The electronic aid which assists the controller will be able to take the local atmosphere conditions into account and set the traffic in such a way that the instructions given to the pilot satisfy the flight envelope.

Flights of modern aircraft are under FMS control. At the present time, FMS may control the flight directly. The main functions are:

a) navigation function: it computes the "best estimate" from the INS, the radio-electric aids such as VOR, DME, TACAN, ILS during final approach and, from now, GPS; it compares the estimated error to the one required over the overflown area. It sends messages such: "NAV ACCY UPGRADED or DOWNGRADED". In the next future, when D-GPS or WAAS will be implemented the correlation will cover all the navigation means.

b) horizontal navigation: this is the type of navigation used today. All the data concerning the way-points (beacon or latitude/longitude coordinates or VOR/VOR, VOR/DME, DME-DME) are stored in the FMS; the drift is computed and from it the horizontal components of the wind vector are computed and delivered to the cockpit. Instantaneous wind vector and actual position (GPS) of the aircraft will be automatically transmitted through an automatic data-link (CNS).

c) vertical profile: the flight envelope is entered into the FMS (or it is read from the Automatic Pilot Computer which needs it); the ATC instructions are injected inside the FMS through a keyboard by the crew. An optimization criterion which operates on a "cost function" tries to determine the optimal vertical profile. This is the "Company Cost Index".

d) extrapolation: the FMS computes the expected times of passage over way points, the expected fuel consumption; from these data the crew is able to forecast any flight plan modification requests which could happen.

3.2.2 Future

It is clear that the FMS, as it stands today, is able to control the flight from initial climb to landing including the braking phase. No major incident has been reported today (i.e. placing the aircraft in a non-recoverable situation)

In the next future, a global optimization, still using the "route concept" will be implemented. The problem will be discussed in paragraph 4.

In the long term future, when the "autonomous flight" will be accepted, the problem of anti-collision is not yet defined. It is hard to believe that T-CAS (or ACAS) will solve the problem when a risk of collision appears. It seems probable that planes will be followed by ATC's and informed by the centers of any possibilities of collisions; to avoid them, initiatives may be let to the pilots or to the controllers on a temporary basis.

3.3 The duality FMS-ATC

In both cases, it will be necessary to "synchronize" the ATC policy - which results from electronic aids - and the instructions injected by the crew into the FMS. This is already an actual problem: when the ATC requests passing over a waypoint, at given *altitude, speed* and *time* the data come from an "optimization process" which tends to reach the maximum rate of landings for example. If the crew injects these data into the on-board FMS, this one computes the instantaneous parameters (speed, vertical velocity...) according to the "Company Cost Index" which, normally is ignored by the ATC. There is no certainty that the extrapolated trajectories computed on board and estimated on the Control Center are the same. Should the crew send the Cost Index used to the ATC or should the FMS send the trajectory computed on board which satisfies the "final conditions" ?

In a shorter way, the problem can be stated as: what is the minimum number of parameters (data) to be exchanged between the ATC and the plane to guarantee that the error between the expected trajectories (ground computed and aircraft computed) does not exceed a given range.

4. Global Optimization Possibility

The "cost index" is specific to a company. At the present time, a local optimization is performed: if a new airplane is entering the sector or if a perturbation has modified the pre-set program, the controller tries to re-arrange the planes in order to be as close as possible of the flight plans (FLP) of the two planes; they ignore the "cost indices" but they assume that the requested flight profile (FLP) minimizes the cost index of each plane.

For the traffic expected in 10 or 15 years, it seems reasonable to look for a "global optimization" that is a way to arrange the planes in such a way that the total penalization of all the planes flying in a zone - which is much larger than a sector - be a minimum.

We suggest to study the feasibility of the following algorithm.[5]

We start from the flight envelope (Fig. 6), more precisely from the flight envelope for a given mass, balance and configuration of the plane. We will consider the cruise phase.

For a given range (say 1000 km) there is an optimum cruise represented by a FL and a Mach number. If the plane is not flying at that point because of ATC constraints, a penalization may be attached to the point representing the flight paramcters. It is easy to imagine "iso-penalization contours" (Fig. 6b) around the optimum point. The point depends in some parameters mass, balance, configuration... of the plane. Algorithms can easily be derived to re-quote the iso-penalization contours around the point which corresponds to the optimum cruise for these parameters. This re-calibration can be made by the ATC computer.

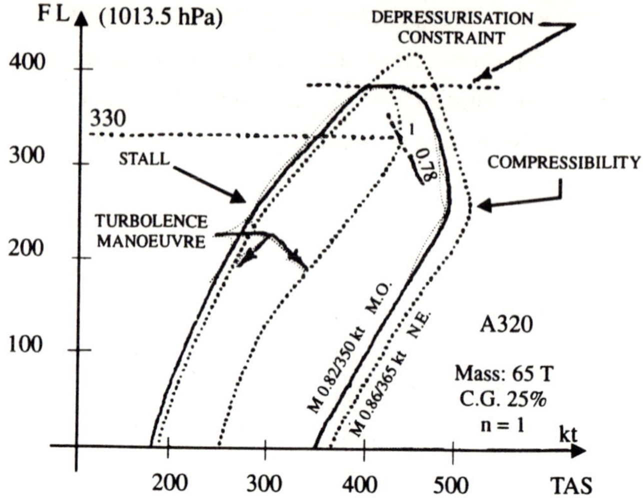

Figure 6a Iso-penalization contours

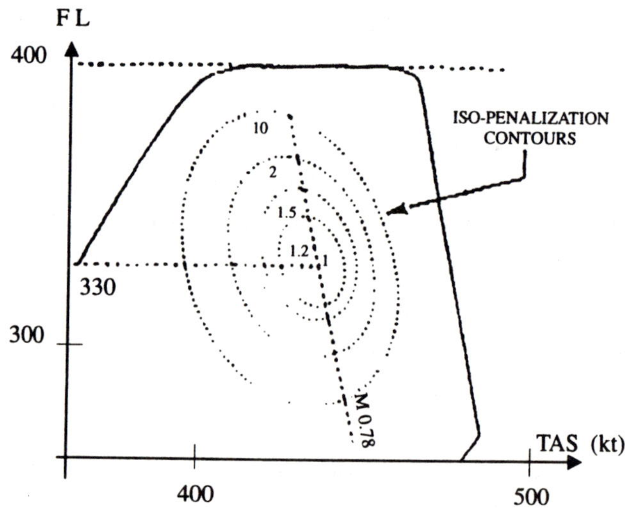

Figure 6b Enlargement around point quoted 1

Let us suppose that, at a given time, the plane arrangement (FL, M) is optimum in a large zone (for example Europe North-West including Holland, Benelux, Allemagne, Suisse, Italie, France, Grande-Bretagne) over a given time horizon (half an hour for example).

The total penalization of all the planes flying in the zone is minimum. Then, an event occurs. It may be:

- the incoming of a plane in the zone
- the outcoming of a plane from the zone
- an error greater than the one accepted for a plane.

A re-optimization process should be engaged: it tries to minimize the total penalization factor for all the planes over a new half an hour horizon. Positively, an event interacts only in a sub-zone, the dimensions of which should be first determined.

If it is simple to state the problem as above, it is clear that many questions should be considered in the final study. For example:

- what about already acquired penalizations, for instance, for a plane coming from abroad ? Do they have to be injected in the minimizing process or should they be ignored ?

- a company may decide to modify the "optimum point" in order to compensate for a delay at takeoff. (A message can be sent to the "global control center " which will use an algorithm to re-quote the penalization contours for this flight).

- it would be advisable to avoid too many changes for a given plane (the Center may determine a collection of solutions closed to the minimorum solution; the final solution could be a compromise between the total penalization and the fact that no planes receives trajectory modification within five minutes after the last one received).

A similar process may be derived for the climb and descent phases. We are conscious of the difficulty of the problem. However, if a global optimization is looked for, only a process based on a general agreement between companies and ATC's could lead to a solution.

5. Conclusions

The continuous growth of the air traffic - above 4.5 % per year - implies immediate actions which improve the present Air Traffic Control as it stands today.

However, looking at an horizon such 2010 or 2015 it seems clear that the "extrapolation " of the actual concept is no longer valid. The problem may be divided into two major spaces: the en-route navigation and the terminal navigation. The boundary between these two spaces is not yet specified.

For the en-route navigation, the "autonomous flight concept" will be accepted and the future FMS's will handle the control of the flight on an optimal basis. The ground control - if it stands - will be informed automatically of the positions of the planes thanks to an automatic reporting system (CNS concept).

For the terminal navigation, which includes approach, landing/takeoff and initial climb, a close coordination between the airplane (crew and FMS) and the ground controllers should be re-enforced. Movements of the planes should be integrated into the "air-segment" in order to comply with a reasonable size of the airport terminals. The air safety factor must be improved - though it has reached an acceptable value today - because of media reactions in case of accident; more and more people will fly: the air safety factor should be aligned on the railway's one though air safety factors are not computed along the same lines : air safety is computed in terms of probability of death per hour of flight (at present 0.5 10^{-7}/h) and ground safety is computed in terms of kilometers run (better than 10^{-9}/megameter run).

References

[1] *The LOTOS panel report DOC.* EUR 1334 EN, January 1994.

[2] E. Petre, *Time-based Air Traffic Control in an Extended Terminal Area,* EUROCONTROL Report 91 2009, June 1993.

[3] *SAATMAS: Fully Automatic Tractors for planes,* IFAC/AS. Workshop on intelligent vehicle, April 1993, Southampton, U.K.

[4] T. Fujita, *The downburs*, University of Chicago, 1985.

[5] M. Pélegrin, *On-line handling of Air Traffic*, AGARDOGRAPH 321 November 1994, "*Towards Global Optimization of Air Traffic Management*".

MULTI-AIRPORT GROUND HOLDING PROBLEM: A HEURISTIC APPROACH BASED ON PRIORITY RULES

Giovanni Andreatta

Università di Padova - Dip. di Matematica Pura ed Applicata
Via Belzoni 7, 35131 Padova - Italy

Lorenzo Brunetta

Politecnico di Milano - Dip. di Elettronica
Piazza L. da Vinci 32, 20133 Milano - Italy

Guglielmo Guastalla

Massachussets Institute of Technology - OR center, E40 -
130 Amherst St. - Cambridge, MA 02139 - USA

In recent years, many sophisticated mathematical models have been proposed to address the problem of reducing congestion at major airports by means of Ground Holding (GH) Policies, i.e., by imposing a ground delay to selected aircraft in order to reduce and possibly to avoid airborne delays and congestion. These models, which are usually optimization models, have found little application in practice so far, probably because they are not easy to understand. In this paper we introduce and analyze a family of simple, easy to understand, heuristics. Each heuristic is based on a specific "priority rule". The flights are grouped dynamically into a manageable number of classes. Each "priority rule" specifies, for any two flights of different classes, which one has priority over the other. The performances of the proposed heuristics are compared among themselves, and with that of an "optimal" policy, on a set of test problems.

1. Introduction

In recent years, many sophisticated mathematical models (see, for instance, [1], [2] or [6]) have been proposed to address the problem of reducing congestion at major airports by means of Ground Holding (GH) Policies, i.e., by imposing a ground delay to selected aircraft in order to reduce and possibly to avoid airborne delays and congestion (see also [4]). These models are usually mathematical optimization models, i.e., they allow the finding of an "optimal" GH policy. "Optimal" means that, under suitable assumptions, the suggested GH policy is the one that minimizes a given penalty function (objective function).

These models however, have found little application in practice so far. A possible explanation is that the penalty function, usually an educated guess of the direct costs of delays to the Airlines, does not necessarily reflect the actual real costs, and even if it does, it is not at all obvious if that is the right criterion for deciding which aircraft a ground delay should be imposed on. Secondly, most of these optimization models are kind of "black boxes" to people who have the responsibility of imposing ground delays, unless they are mathematicians or Operations Research experts, which is almost never the case.

As a matter of fact, so far, the FAA has preferred to use Simulation models rather than Optimization models for implementing its GH Policy. The reason is, we believe, that in a Simulation model it is much easier to understand what is going on and therefore, from a psychological point of view, it is much easier to trust these models rather than to rely on somehow obscure Optimization models.

In this paper we will introduce and analyze a family of simple, easy to understand, heuristics. Each heuristic is based on a specific "priority rule". First, the flights are grouped dynamically into a manageable number of classes. In the following examples, each class is characterized by the amount of delay already suffered by a flight and by how many flights are later connected to the current one. Each "priority rule" specifies, for any two flights of different classes, which one has priority over the other.

We should make clear at this point that the focus of the paper is not to suggest the use of any particular priority table. It should be the Expert, i.e., the person in charge of making the actual GH decisions and having the authority to enforce them, who should choose a priority table he feels confident and comfortable with. Such a priority table can always be elicited from the Expert if it is not directly available. We assume here that a priority table is given.

There is a second reason for investigating these or other simple heuristics. Usually, to achieve a solution, an optimization algorithm requires a computational time that is much longer than that required by a heuristic one and since the computational time is an important factor for evaluating a GH algorithm, it is worthwhile to explore possible heuristics.

Of course, for a heuristic to be recommended, it is not enough to be simple and easy to compute. It is also necessary to be "good", i.e., to perform reasonably well when compared with the absolute best in terms of a specified criterion. In the sequel we will also discuss the performance of the proposed heuristics. We will compare them among themselves and with an "optimal" policy on a set of test problems. The criterion chosen is the objective function used in [1], [2], [5] and [6], which is directly proportional (by a factor of 50) to the cumulative amount of delay imposed on all flights.

We conclude this section by addressing the new emerging concept of "free flight". This concept is born among the Aircarriers of the United States that are pushing the FAA to provide them only with the arrival timeslot, leaving the Airlines the "freedom" of departing, of selecting the routes and speeds, as long as they are ready to land at the assigned timeslot. More rigorously – borrowing the words of J. Fearnsides from his presentation at the Conference Air Traffic Management '95 held in October 1995 in Capri – free flight is "a safe and efficient flight operating capability under instrument flight rule in which the operators have the freedom to select their path and speed in real time. Air traffic restrictions are only imposed to ensure separation, to preclude exceeding airport capacity, to prevent unauthorized flight through special use airspace, and to ensure safety of flight. Restrictions are limited in extent and duration to correct the identified problem." The heuristics we present in the paper can easily fit into the "free flight" concept. In this case, the "delays" suggested by the heuristics have to be interpreted as time differences between the arrival time slot imposed by the FAA, or other analogous Authority, (leaving up to the Aircarriers to decide how much delay should be absorbed on the ground, how much through speed reduction or through rerouting, etc.) and the time slot originally scheduled. Under a "ground holding policy" all planned delays are assumed to be incurred on the ground prior to departure.

2. The Algorithm

Before presenting the algorithm, let us state explicitly the following assumptions:

A1) congestion may only arise in proximity of arrival airports and there is no congestion in sectors nor in the departure airports,

A2) each airport landing capacity has a time profile which is considered to be known and deterministic,

A3) there are consecutive connected flights.

While the first hypothesis is not true in actual reality, it can however be considered as a first, reasonable approximation, particularly in the USA. The second hypothesis reflects current standard practices. The third hypothesis has been made only to avoid trivial instances of the problem.

The heuristic algorithm is iterative and may be outlined as follows:

i) for each time slot and for each airport create a list of all flights candidate to arrive to that airport at that time slot;

ii) order the flights in the list according to decreasing priorities (based on a given priority table);

iii) if the landing capacity at a given airport during a given time period is K, assign the first K flights in the list to arrive during that time slot and correspondingly compute their (ground holding) delays. The remaining flights in the list (if any) become candidate for the next time period.

We can now introduce the notations we will use in the following pages.

T	=	set of time intervals;
t	=	generic time interval ($t=1,\ldots,\ \lvert T\rvert$);
Z	=	set of airports;
z	=	generic airport ($z=1,\ldots,\ \lvert Z\rvert$);
F	=	set of flights;
f	=	generic flight ($f=1,\ldots,\ \lvert F\rvert$);
$K_{z,t}$	=	capacity of airport z at time t , for all $t \in T$ and for all $z \in Z$;
$succ_f$	=	the flight that is the immediate successor of flight f (if f is the last leg of a trip, we set $succ_f = 0$), for all $f \in F$;
$pred_f$	=	the flight that is the immediate predecessor of flight f (if f is the first leg of a trip, we set $pred_f = 0$), for all $f \in F$;
r_f	=	desired arrival time for flight f, for all $f \in F$;
a_f	=	actual arrival time for flight f, for all $f \in F$;
e_f	=	earliest arrival time for flight f, compatible with the actual arrival time of $pred_f$, for all $f \in F$;
l_f	=	latest arrival time interval for flight f, compatible with the desired arrival time of $succ_f$, for all $f \in F$;
$s_{f,succ_f}$	=	maximum delay for flight f without causing any delay on flight $succ_f$;
Δ_f	=	delay suffered by flight f, for all $f \in F$;
Δ_{max}	=	maximum allowed number of delay periods per flight;
π_{f,Δ_f}	=	priority index assigned to flight f, when it has already suffered a delay Δ_f, for all f in F and $\Delta_f \in \{0,\ldots,\ \Delta_{max}\}$;
prt_f	=	current priority index assigned to flight f, for all $f \in F$;
Π	=	$[\pi_{f,\Delta_f}]$ = a given "priority table".

Notice that $s_{f,succ_f}$ is the maximum delay that flight f may suffer without causing any delay on the successive connected flight $succ_f$ this is given by the difference between the departure time of flight $succ_f$ and the arrival time of flight f minus the time needed for cleaning, refueling, unloading and loading passengers, etc.

In the following Procedure ABG all flights receive a label E, W or A . Label E means that a flight is currently in the examination list; label W means that a flight is waiting (to enter in one of the examination lists) and label A means that a flight is assigned to arrive.

Procedure ABG. *(Heuristic for the multi-airport ground holding problem)*

Input: Π, t, Z, F, $K_{z,t}$ for all $t \in T$ and for all $z \in Z$, F_z for all $z \in Z$, r_f, $pred_f$, $succ_f$ for all $f \in F$.

Step 0. *Initialization.* Set $t = 1$, $a_f = \infty$, $e_f = r_f$, label(f) $= E$ if $pred_f = 0$ and label(f) = W if $pred_f > 0$, for all $f \in F$.

Step 1. *Creating examination lists.* For every $z = 1, \ldots, |Z|$, *let* $F_{z,t}$ be the list of flights f such that label(f) $= E$ and $e_f \leq$ t, ordered according to decreasing priorities $prt_f = \pi_{f,t-r_f}$.
Set $z = 1$.

Step 2. If $K_{z,t} = 0$ or $F_{z,t} = \emptyset$, then go to Step 4, else let f be the first flight in $F_{z,t}$.

Step 3. *Flight f arrives at time t.* Set label $(f) = A$, $a_f = t$, $K_{z,t} = K_{z,t} - 1$, $F_{z,t} = F_{z,t} \setminus \{f\}$.
If $succ_f > 0$ then label($succ_f$) $= E$ and $e_{succ_f} = r_{succ_f} + \text{MAX}\{0, a_f - r_f - s_{f,succ_f}\}$.

Step 4. If $z < |Z|$ then set $z = z+1$ and go to Step 2.

Step 5. If $t < |T|$ then set $t = t+1$ and go to Step 1.

Step 6. For all $f \in F$, let $\Delta_f = a_f - r_f$. Stop.

Output: Δ_f for all $f \in F$.

In Step 1, examination lists, based upon a priority table given in input, are created and updated. In Step 2, whenever possible, a flight f is selected to arrive. In Step 3 we update whatever is affected by the arrival of flight f at time t. Step 4, Step 5 and Step 6 have obvious interpretation. By providing different priority rules to Procedure ABG, one obtains different heuristics, with different performances. Some of these are presented in the next section.

3. A Selection of Priority Rules

A "Priority Rule" is specified by a "Priority Table", i.e., a matrix with as many columns as there are flights and as many rows as the possible number of delay periods, which assigns to every flight *f*, that has suffered a delay Δ_f, a priority index $\pi_{f,\Delta f}$. These priority indexes are then used in the algorithm for scheduling the departure of the various flights.

Since it is conceivable that some flights share the same priority index, it is convenient to group them into equivalence classes. This drastically reduces the amount of memory required to store the priority table. In this paper we choose to aggregate into the same class all flights that are followed by the same given number of connected flights. Of course, other aggregations are possible.

We modeled our priority rules on some of the test problems first proposed in [5], and then used in [1], [2], [3], and [6]. There, the maximum allowed delay for a single flight *f* is 4, i.e., $\Delta_{max} = 4$, and the same aircraft might perform a trip of at most 5 legs. Furthermore $s_{f,succ_f}$ is a constant for every pair of consecutive flights and it is equal to 1. Procedure ABG does not guarantee to provide a solution that will automatically satisfy the constraints $\Delta_f \leq \Delta_{max}$ for all $f \in F$. Therefore the proposed solution could be unfeasible. This depends on the choice of the priority table. For instance, of the 5 priority tables introduced in this section, only the first 3 did always supply feasible solutions. These considerations led us to the construction of various priority tables, some of which are reported next. In tables A and B we divide the flights into two classes, flights with no successor (*no_succ*) and flights with (one or more) successors (*w_succ*). In table A we simply give high priority to flights with more units of delay and with successors.

	no_succ	w_succ
$\Delta_f = 0$	0	1
$\Delta_f = 1$	2	3
$\Delta_f = 2$	4	5
$\Delta_f = 3$	6	7
$\Delta_f = 4$	8	8

Table 1. Priority table A

In priority table B the flights that have already a delay of 4 time units have the highest priority, followed by those that have already a delay of 3 time units. Notice that we give higher priority to flights with successors and without delays than to flights with no successors and 1 or 2 units of delay.

	no_succ	w_succ
$\Delta_f = 0$	0	3
$\Delta_f = 1$	1	4
$\Delta_f = 2$	2	5
$\Delta_f = 3$	6	7
$\Delta_f = 4$	8	8

Table 2. Priority table B

In tables C and D we divide the flights into five classes depending on how many legs of the same trip follow flight f.

In table C the strategy is the same as in table B for the first two columns; we increase the priority of a flight depending on the number of its successively connected flights. In the third, fourth and fifth columns we give higher priority to flights with a larger number of successors rather than to flights with a larger delay. For example, we schedule the departure of a flight with 4 successive flights and 2 units of delay before that of a flight with 2 successors that has already suffered a delay of 3 units.

	0_succ	1_succ	2_succ	3_succ	4_succ
$\Delta_f = 0$	0	30	33	35	37
$\Delta_f = 1$	10	40	43	45	47
$\Delta_f = 2$	20	50	65	75	85
$\Delta_f = 3$	60	70	80	90	100
$\Delta_f = 4$	110	110	110	110	110

Table 3. Priority table C

Table D is constructed considering the total delay produced by delaying a flight for one further period of time. For example, consider a flight with four successive flights that has already suffered 2 units of delay. A delay of 3 units on the first leg of a five legs trip automatically implies a delay of 2 units on the second leg, 1 unit on the third leg and no delay on the fourth and fifth legs.

	0_succ	1_succ	2_succ	3_succ	4_succ
$\Delta_f = 0$	1	1	1	1	1
$\Delta_f = 1$	1	2	2	2	2
$\Delta_f = 2$	1	2	3	3	3
$\Delta_f = 3$	1	2	3	4	4
$\Delta_f = 4$	1	2	3	4	5

Table 4. Priority table D

Finally, there is a "passive" strategy (P), that assigns to every flight the same priority, i.e., $\pi_{f,\Delta f} = 1$.

4. Computational Results

The algorithm described in the previous sections was implemented in C on a SUN Sparc 10/41 with 64 Mb of RAM and Solaris 2.3 as operating system; the performance of the code was tested on some of the multi-airport ground holding problems used by P.B. Vranas in his Ph.D. thesis [5] that were gently provided to us by A.R. Odoni. They are the same instances used by P.B. Vranas, D.J. Bertsimas, A.R. Odoni in [6], by Andreatta and Tidona in [2] and by Andreatta and Brunetta in [1] to test their integer programming formulations. The computation time to solve any of the seven test cases is always less than 2.4 seconds on the Sparc 10/41. By contrast, see [1, 2, 3, 5 and 6], to obtain the optimal solution by means of an exact procedure, required, for some of the considered problems, more than 1 hour.

In table 5 we recall for each test case, some of its data. Namely, K (the number of airports), F (the total number of flights), $K(z) = K_{z,t}$ (the capacity of airport z, which happens to be independent from t), and *i-legs* (the number of trips composed exactly of i legs).

Test	*K*	*F*	*K(1)*	*K(2)*	*K(3)*	*K(4)*	*1-leg*	*2-leg*	*3-leg*	*4-leg*	*5-leg*
1	2	1000	12	14			600	200			
2	2	1000	10	10			200	400			
3	2	1000	11	11			100		300		
4	2	1000	10	10							200
5	4	2000	14	14	14	14	1200	400			
6	4	2000	14	14	14	14	500	600	100		
7	4	2000	12	12	12	12	200	300	100	100	100

Table 5. Problems data

In tables 6 and 7 we compare the performance of the various algorithms. In table 6, the selected criterion is the same used in [6] and [2], i.e., the cumulative amount of delay imposed on all flights. For every test problem we report the optimal value (column OPT) of the objective function and for each of the priority tables ($\Pi = A$, $\Pi = B$, ..., $\Pi = P$) we report the corresponding value of the objective function.

Test	OPT	Π =A	Π =B	Π =C	Π =D	Π =P
1	71000	71000	71000	71000	71000	71000
2	56000	56000	56000	56000	56000	56000
3	84700	84800	84800	84800	84200	95950
4	65000	65000	65000	65000	65000	65000
5	96300	109400	96300	96300	96300	96300
6	92200	99300	96500	96500	88300	88300
7	72200	86000	72800	72800	75800	113800

Table 6. Objective function value using different priority-tables with procedure ABG

In Table 7 we report the maximum number of delay periods imposed on a single flight by an optimal policy (column Δ_{OPT}) and the one imposed when adopting priority tables A, B, ..., P. From this table one can see that priority tables A, B, C provided always a feasible solution, whereas priority tables D and P never did.

Test	Δ_{OPT}	Δ_A	Δ_B	Δ_C	Δ_D	Δ_P
1	4	4	4	4	6	6
2	3	3	3	3	7	7
3	4	4	4	4	12	11
4	4	3	4	4	6	13
5	4	4	4	4	13	13
6	4	4	4	4	13	13
7	4	4	4	4	5	7

Table 7. Maximum number of delay periods using different priority tables with procedure ABG

Priority rule A, which could seem the most natural, allows to solve at optimality only 3 problems out of 7. In any case it always provides a feasible solution. Priority rule B allows to solve at optimality 4 out of 7 problems, again providing a feasible solution for all of them. The highest gap occurs in problem 6, where our algorithm gives a solution that is 4.5 % worse than the optimal one, while in the other two cases is worse by less than 1 %. Priority rule C gives the same results as B, showing that we do not obtain any improvement by augmenting the number of columns in the priority table, i.e., dividing the flights into more categories.

Priority rule D provides a lower objective function for the problems that were not solved at optimality with strategies B and C, but the solutions are never feasible under the requirement that Δ_{max} should be lower or equal to 4. Also notice that in test cases no. 3 and 6 the values of the objective function evaluated as the (unfeasible) solutions obtained adopting priority table D are "better" than the optimal ones.

5. Improved Algorithm

The Algorithm ABG proposed in the previous Sections is a constructive procedure that provides a schedule which is not necessarily optimal. As it is typical with scheduling or sequencing heuristics, one can hope to improve upon the given schedule by trying to operate an exchange of arrival times between flights.

In this Section we describe a modified algorithm, where a tentative exchange is operated within the constructive procedure. The resulting algorithm, let us call it MABG, is identical to procedure ABG except for Step 0 (where a new instruction is added) and Step 3 (which is substituted by the new Steps 3.1, 3.2 and 3.3).

In Step 0 of MABG, the following instruction is added to those in Step 0 of ABG:

Set: $l_f = r_f + s_{f,succ_f}$ if $succ_f > 0$ and $l_f = r_f + \Delta_{max}$ if $succ_f = 0$.

Step 3 in ABG is substituted by the following:

Step 3.1. *Flight f is assigned to arrive.* Set label(f) = A, $a_f = t$, $K_{z,t} = K_{z,t}-1$, $F_{z,t} = F_{z,t} \setminus \{f\}$.
If $succ_f = 0$ then go to Step 4. If $t - r_f \leq s_{f,succ_f}$ then go to Step 3.3.

Step 3.2. *Tentative exchange.* Let f be the first flight with the lowest e_f among the flights f' such that $e_{f'} \leq a_{f'} < t$ and $t \leq l_{f'}$. If no such f exists, then go toStep 3.3, else set $a_f = a_{f'}$, $a_{f'} = t$ and go to Step 3.3.

Step 3.3. *Updating the successor of f.* Set label($succ_f$) = E and $e_{succ_f} = r_{succ_f} + \text{MAX}\,\{0, a_f - r_f - s_{f,succ_f}\}$.
Updating the predecessor of f. Set $l_{pred_f} = \min\{l_{pred_f} + \Delta_f, r_{pred_f} + \Delta_{max}\}$.

The key idea beneath procedure MABG is to try to exchange the arrival times of a flight f, which will cause an automatic delay on a successive flight, with the arrival time of an already scheduled flight f, that will not cause such a problem. The algorithm MABG was implemented in C (on the same hardware as ABG) and tested upon the same problems described in Section 4. The results are presented in table 8 and 9, using the same priority rules as in the previous Section. The computational time required to run MABG was comparable with that for ABG: again each problem was solved in less than 2.4 seconds.

Test	OPT	Π =A	Π =B	Π =C	Π =D	Π =P
1	71000	71000	71000	71000	71000	71000
2	56000	56000	56000	56000	56000	56000
3	84700	84700	84700	84700	84200	95950
4	65000	65000	65000	65000	65000	81500
5	96300	97300	96300	96300	96300	96300
6	92200	94400	94400	94400	88300	88300
7	72200	86000	72800	72800	75800	97000

Table 8. Objective function values using different priority-tables with procedure MABG

Test	Δ_{OPT}	Δ_A	Δ_B	Δ_C	Δ_D	Δ_P
1	4	4	4	4	6	6
2	4	4	4	4	7	7
3	4	4	4	4	12	11
4	4	3	4	4	6	13
5	4	4	4	4	13	13
6	4	4	4	4	13	13
7	4	4	4	4	5	6

Table 9. Maximum number of delay periods using different priority tables with procedure MABG

The most important remarks that should be made are the following:

i) Priority rule C continues to provide the same results, in terms of objective function value and maximum number of delay periods, as priority rule B;

ii) When adopting priority rule B or C, MABG solves to optimality 5 problems out of 7 (4 were solved to optimality by ABG) and, on problem 6, it improves the objective function value;

iii) When adopting priority rule A, MABG provides almost always the same results in terms of objective function value (with the exception of problem 5) and maximum number of delay periods, as priority rule B or C. Furthermore, MABG substantially improves, with respect to ABG, the performance of priority rule A;

iv) There is almost no improvement (with only one exception) in terms of maximum number of delay periods between running MABG or ABG;

v) Priority rules D or P continue to perform very poorly.

6. The BGN Test Cases

We used the Brunetta, Guastalla and Navazio test cases (BGN test cases) to test the quality of the algorithms on a set of larger and more complex instances. The daily timetable of the flights of the new instances was generated by means of the Pseudo Official Air Guide Generator (POAGG) of the Draper Laboratory. Airport capacity range from 1 to 5 arrivals for each time period for small airports, and 10 to 14 arrivals for the others. These cases will be referred in the sequel as the test cases 8 to 39 (BGN test cases). A concise description of each instance (the number of airports, its total number of flights and the number of flights having a given number of legs) is reported in Table 10. The set of 32 instances can be accessed by other researchers interested in developing and testing the quality of their exact and heuristic algorithms via anonymous ftp at the following site: *ftp.math.unipd.it*. A complete description of BGN test cases is provided in [3].

Test	Z	F	leg	2legs	3legs	4legs	5legs	6legs	7legs	8legs	9legs	10legs	11legs
8	2	1004	50	431	24	1				2			
9	3	1172	77	119	114	30	29	12	17	5	1	1	
10	3	1181	77	124	100	47	27	13	9	8	2	1	
11	2	1342	67	71	130	49	40	26	11	6	5	1	1
12	9	1403	70	525	52	18	3	4		2			
13	9	1419	76	514	73	20	2	1					
14	3	1593	120	139	148	62	36	22	7	7	6	1	2
15	5	1760	23	67	65	56	57	35	26	46	13		2
16	4	1854	86	89	129	66	46	37	29	17	14		2
17	2	1909	90	784	73	4				2			
18	4	1940	93	291	184	71	32	21	10	8	1		
19	3	1945	66	470	199	49	20	5		2			
20	3	1989	124	192	179	69	42	39	13	6	7		2
21	4	2366	98	360	217	88	57	22	8	9			
22	9	2396	130	735	128	52	28	8		2			
23	3	2526	127	1023	81	15	2	4		2			
24	3	2527	134	1006	94	11	3	4		2			
25	3	2530	132	1010	90	12	4	4		2			
26	3	2532	126	1020	86	12	4	4		2			
27	3	2534	128	989	120	17							
28	6	2546	147	464	243	83	31	25	9	3	2		
29	4	2672	107	915	199	29	3		1				
30	4	2806	95	612	260	90	36	9	7	8			
31	5	2882	44	148	132	114	60	60	52	28	41	4	3
32	6	3034	109	93	184	95	65	90	59	32	27	3	
33	10	3142	197	902	201	70	42	8					
34	5	3192	63	99	183	122	75	75	39	48	40	3	2
35	5	3805	185	700	366	134	55	20	17	9			
36	5	3823	149	144	276	148	118	79	50	38	22	5	
37	6	4523	227	549	393	173	118	56	34	17	3		
38	5	4773	240	921	433	153	79	36	15	8			
39	10	5005	11	1481	366	142	56	12	2				

Table 10. The BGN Problems data

Test	BS	MABG	Δ_C	Err (%)
8	13200	13200	4	0
9	51000	52850	4	3.63
10	26150	26300	4	0.76
11	51500	52350	4	1.65
12	11250	11250	4	0
13	10750	10750	4	0
14	58900	60300	4	2.37
15	62400	63150	4	1.20
16	79350	81000	4	2.08
17	42700	43050	4	0.82
18	54750	55350	4	1.10
19	22400	22400	4	0
20	45150	45250	4	0.055
21	63550	64150	4	0.94
22	70500	71300	4	1.13
23	52150	52400	4	0.48
24	25250	25250	4	0
25	27000	27050	4	0.19
26	32600	32800	4	0.61
27	39800	39900	4	0.25
28	74250	77850	4	4.85
29	43900	44300	4	0.91
30	75300	76650	4	1.79
31	138350	141400	4	2.20
32	146400	150250	4	2.63
33	80950	81850	4	1.11
34	189600	193150	4	1.87
35	113600	115700	4	1.85
36	124000	125850	4	1.49
37	137800	140850	4	2.21
38	172450	176900	4	1.02
39	134500	140100	4	4.16

Table 11. Statistics of the MABG algorithm on the BGN test cases

The BGN instances were solved to optimality by the BS model (see [1], or [3]). For each test case we report, on table 11, statistics for the MABG procedure (with priority table C): the optimal value of the objective function (BS) obtained by using the BS model, the heuristic objective function value (MABG), the maximum number of delay periods imposed on a single flight by the MABG (Δ_C), and the percentage error between the two objective function values (Err (%)).

The MABG algorithm solves at optimality only five instances out of 32, but the percentage error in the worst case is 4.16 %. Using the BS model it takes almost two hours to solve instance 37 on the SUN Sparc 10/41, while the MABG algorithm requires less than six second to solve each one of the 32 BGN instances. The maximum number of delay periods is 4, which means that all the solutions provided by MABG are feasible.

7. Concluding Remarks

We think that the preceding sections have gathered enough evidence about the good properties of the heuristic algorithms we are suggesting and that can be summarized as follows. The proposed heuristic are:

i) simple to understand;
ii) flexible: properly choosing a priority table can produce a ground holding policy that fits particular needs;
iii) easy to implement;
iv) very fast: few seconds of CPU are required;
v) reasonably good: a proper choice of the priority table produces results within 5% of the “optimal”(according to a specific criterion) ones.

Acknowledgements

The research of the first author was partially supported by two research grants from *Progetto Finalizzato Trasporti 2* of the Italian National Research Council (contracts CNR/93.01778.PF74, CNR/94.01335.PF74). We would like to thank Amedeo Odoni for his valuable suggestions and encouragements during our research. We are also grateful to Lisa Navazio for her help in implementing the MABG algorithm and for providing us the BGN test cases.

References

[1] G. Andreatta and L. Brunetta. Multi-airport ground holding problem: a computational evaluation of exact algorithms, to appear in *Operations Research*, 1996.

[2] G. Andreatta and G. Tidona. *A new formulation for the multi-airport ground holding problem.* Internal Report no. 3/94, Dip. di Matematica Pura e Applicata, Università di Padova, Italy, March 1994.

[3] L. Brunetta, G. Guastalla and L. Navazio. *A new approach for solving the multi-airport ground holding problem.* Internal Report no. 25/95, Dip. di Matematica Pura e Applicata, Università di Padova, Italy, September 1995.

[4] A.R. Odoni. The flow management problem in air traffic control, in A.R. Odoni, L. Bianco and G. Szego eds., *Flow control of congested networks*, Springer Verlag, Berlin, 1987.

[5] P.B. Vranas. *The multi-airport ground holding problem in air traffic control,* Ph.D. Thesis, Dept. of Ocean Engineering and Operations Research Center, MIT, Cambridge, Massachusetts, 1993.

[6] P.B. Vranas, D.J. Bertsimas and A.R. Odoni. The multi-airport ground holding problem in air traffic control. *Operations Research,* 42:249-261, 1994.

NETWORK OPTIMIZATION IN AIR TRAFFIC MANAGEMENT

Mario Lucertini, Stefano Smriglio

Centro Vito Volterra, University of Rome "Tor Vergata"
Viale della Ricerca Scientifica – 00133 Rome, Italy

Daniela Telmon

TRADEOFF, consulenza e servizi per le aziende, Roma

Many problems in air traffic management have a common aspect: it is required to minimize the lead time of goods and passengers through the system. Each of these elements, while passing through the system, must undergo a given sequence of operations. Operations management procedures in traffic systems are generally based on a planning phase and on a real time one. In the former, items are allocated according to capacity and other regulations (e.g. slot allocation). The latter is based on the idea of pushing items through the system every time this is possible and thus shifting downstream the solution of the conflicts that may happen in the following phases of the operation chain. This looks very similar to the well known push strategies in manufacturing. Nevertheless, in manufacturing environment, pull strategies have often proved to be more effective. On this basis, in this paper, we represent the Flight Information Region (FIR) as a service network, and propose some experiments to verify the effect of pull type strategies in real-time flow management.

1. Introduction

The design of an air traffic management (ATM) system is mainly focused on the dynamic control of congestion problems. The evaluation and the optimization of the performances are based on few basic concepts: system capacity or throughput, lead time and/or work in progress, productivity and cost-effectiveness, flexibility, and complexity.

In the current air transportation network airports are recognized to be important elements for the definition of global system's capacity. The airport is often a bottleneck for both aircraft flow management (FM), and for passenger and luggage handling. As far as the former is concerned, the airport, in many cases, cannot perform as many movements (take-offs and landings) in a given time interval as required, because of aircraft flow management constraints, and it produces undesired delays and costs. Among the reasons that can be quoted for this limited capability there are the rapid growth of customer demand, the difficulty of increasing system's capacity by building new airport facilities or enlarging the existing ones, and also the lack of involvement of adequate analytical and simulation tools in FM [5].

Starting from Odoni [6], where the guidelines of the problem have been stated, a lot of research has been done in the FM field. Bianco an Bielli [5] claim that the problem can be faced at each phase of the planning process, while the main part of the optimization models they propose refer to the slot allocation problem, arising in tactical planning. In particular, the effectiveness of *Ground-Holding* policies has been widely investigated by many authors [4, 13].

It can be agreed that a set of methodologies, management procedures and integration strategies, outlined for other purposes (like operations management in manufacturing systems or in the service sector, task management in distributed processing systems [1, 3, 7]) may be relevant for improving performance in air transportation system [11]. Of course, the different techniques may have to be modified or redesigned to satisfy particular needs, but the first experiments, undertaken in different countries on the basis of simulation, look promising [8, 10].

The aim of our work is to introduce new strategies for real-time flow management inside a terminal control region, on the basis of experiences performed in manufacturing systems. The idea is based on the following considerations.

Operations management procedures in traffic systems are generally based on a planning phase, where aircraft are allocated according to capacity and other regulations (e.g. slot allocation), and on a real time phase. The latter is generally performed by pushing aircraft through the system every time it is possible, shifting downstream the solution of conflicts that may happen in following stages of the operation sequence. This looks very similar to the well known push strategies in manufacturing.

On the other hand these concepts have proved to be less efficient than other kind of strategies (e.g. pull-type) in many modern manufacturing systems.

It is interesting to note that some of these concepts have already been experimented in ATM. For example, the ground holding policies [4, 13] can be interpreted as pull type management strategies, based on the idea that a plane is allowed to leave the parking space only when all its way from there on is clear, and it can reach destination without further delays (waiting in the parking space is obviously less expensive than waiting in further ground positions or in the airspace).

2. Measuring Performances

A preliminary step for measuring performances is to find a suitable representation of the system structure and functions.

Generally, the most effective way to represent service systems, like the ATM, one is the so called *process* approach. Following this methodology, we must identify users of the system (input from outside) and the sequence of operations that each user must undergo (in the different situations which can occur during the time interval considered for the analysis) to complete its service, producing an output.

The operations performed on all users are subdivided into several linked elementary processes, so that similar operations performed at equivalent stages of the operation chain on different users, are possibly clustered in an elementary process. Each elementary process is characterized by five elements: the operations performed, the format of the input, the format of the output, the upstream processes and the downstream processes. To build up a measuring system we must evaluate all the above five elements. The connection with upstream and downstream processes produce the so called *network of processes*, which represents the functions of the whole service system.

To perform services, resources are consumed or kept busy. The flows of resources (materials, information, machines, ...) are embedded into the network of processes (often structured in a resource network), and feed the elementary processes at the right time and in the right way.

The behaviour of the flows in the network of processes is driven by a decision making system which produces the information needed to start-up the operations and to circulate correctly the flows of resources and users. With such representation it is generally possible to introduce a measuring system to extract many information useful for the decision making process and for performance evaluation purposes.

In practice, the measures (time, quantity, cost, quality, ...) directly obtained from the real system are generally not enough for producing good solutions and a suitable evaluation. We need also measures obtained performing experiments with a conceptual model (often in form of a simulation model) on the ground of artificial scenarios. All the measures can feed the set of performance indicators chosen to represent the system behaviour. From the numerical values of the indicators we can produce an acceptable evaluation of performances, by quantifying an objective function (or, in a more realistic situation, a set of objective functions), a benchmark and a rating system. On the ground of the objective function, the benchmark, the rating system and the value of the indicators, we can outline a procedure for evaluating the performances. Such an evaluation becomes an input for the decision making system (and, therefore, for performances improvement strategies) and for requirement assessment procedures or performances comparison.

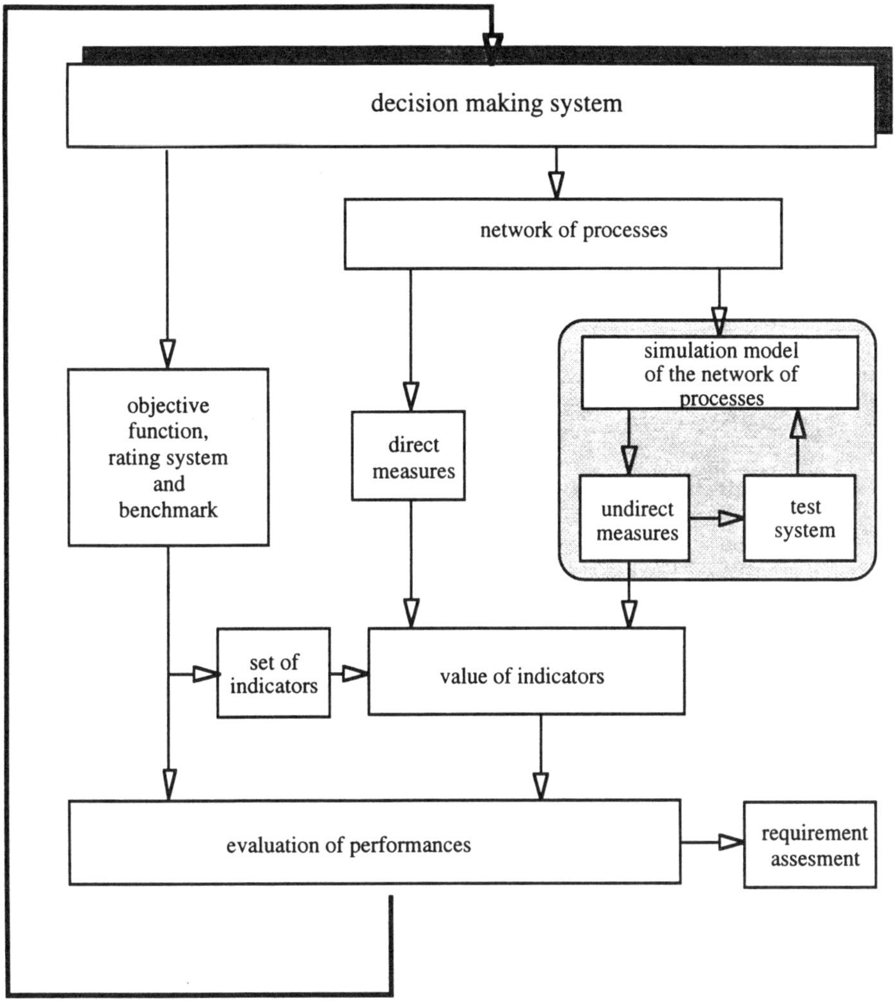

Figure 1. The Performance Evaluation System

The scheme of the process of performance evaluation is represented in Figure 1. We must remark that the test system depends on the objective function, thc rating system and the benchmark, and it also affects the set of indicators.

3. Pull and Push Strategies in Service Networks

In production systems research, several studies concern *push* and *pull* strategies for material flow management in manufacturing lines. The same concepts can be properly applied to those service networks whose structure is similar to a flow line. In ATM it is possible to identify some processes having this feature.

Let now introduce a formal definition of push and pull systems in service networks. To point out the model we will follow the development presented in [12].

Consider a service system consisting of a sequence of n single server processing facilities (service stations), denoted as $j=1,2, \ldots, n$. Consider a set of m users, denoted as $i=1,2, \ldots,m$. Each user must be processed by all the stations in a definite order, which is the same for all users. Assume that the machine names $j=1,2, \ldots, n$ are ordered according to this order (Figure 2). In the following we will define the processing (service) of a user i by a station j as an *operation,* denoted as $o(i,j)$.

Each user is characterized by its release time $a(i)$, representing the time instant at which the user i can start its service at the first station, and its due date $b(i)$, which is the time instant within which it is expected to leave the last machine.

Each pair (i,j) user-station is characterized by the service time $d(i,j)$, representing the duration of the operation $o(i,j)$.

Assume that there are n buffers (one before each station), denoted as $j=1,2, \ldots,n$ in the same order of machines. A buffer j can accommodate users waiting for operation $o(i,j)$ (Figure 2). Each buffer j has a capacity $k(j)$, expressing the maximum number of users it can contain in any given moment. Two particular cases are zero and very large (theoretically infinite) values for $k(j)$. We assume buffer 1 to have always infinite capacity.

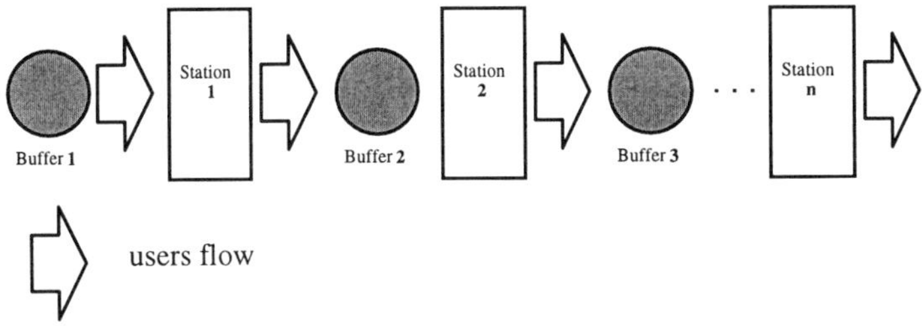

Figure 2. Sequence of Service Facilities

In order to characterize the state of the single element of the system, it is worthwhile to introduce some more definitions. We say that a service facility at a given time instant is *busy* (*idle*) when there is (there is not) a user inside of it. Note that when a station is busy there is a user either receiving the service (i.e. an operation is running) or waiting for leaving the station.

A buffer at a given time instant is said to be *full* (*not-full*) if it accommodates a number of users equal to (strictly less than) its capacity.

A user, at a given time, is said to be *processed* if it is currently receiving the service, otherwise it is defined *waiting*.

Note that a waiting user can be either in a buffer or inside a service station, still waiting for processing, or after having received the service. In order to complete the definition of the system, we specify some more operating rules:

- the sequence in which users visit each station is the same for all the stations. Each buffer is managed on the basis of a FIFO discipline. We assume, with no loss of generality, that the names of users $i=1,2, \ldots, m$ are ordered according to this sequence.
- each user is an indivisible unit, i.e. it cannot be shared by different service stations;
- each station can serve at most one user at a time;
- for each pair (i, j) user-station, once operation $o(i,j)$ has started, it cannot be interrupted until it has finished;
- each operation $o(i,j)$ can start just after the user $i-1$ exits the station j, i.e. no set-up procedures are required for the stations.

In order to represent the system's working, we now introduce a linear programming model. Define the decision variables:

- $x(i,j)$, time instant at which user i enters the station j;
- $y(i,j)$, time instant at which user i leaves the station j.

The problem, which in the following will be referred as Services Planning (SP), is to find $x,y \in \Re^{m+n}$ such that:

(i)	$x(i,1) \geq a(i)$,	$i = 1,2, \ldots, m$;
(ii)	$y(i,n) \leq b(i)$,	$i = 1,2, \ldots, m$;
(iii)	$x(i,j) \geq y(i-1,j)$,	$i = 1,2, \ldots, m, \; j = 1,2, \ldots, n$;
(iv)	$x(i,j) \geq y(i,j-1)$,	$i = 1,2, \ldots, m, \; j = 1,2, \ldots, n$;
(v)	$y(i,j) \geq x(i,j) + d(i,j)$,	$i = 1,2, \ldots, m, \; j = 1,2, \ldots, n$;
(vi)	$y(i,j) \geq x(i-k(j+1),j+1)$,	$i = k(j+1)+1, k(j+1)+2, \ldots, m$, $j = 1,2,\ldots,n$.

The constraints *(iii)* state that each user (except the first one) cannot enter a station before the previous user has left it. Constraints *(iv)* state that each user cannot enter a station (except the first one) before it has left the previous station. Constraints *(vi)* come from the limited buffers capacities and from FIFO discipline: each user cannot leave a station j if the output buffer $j+1$ has no empty positions.

Any feasible solution of inequalities *(i)*-*(vi)* represents a possible way the service system can perform all the operations. We now describe two different solution methods for (SP), corresponding to the well known push and pull strategies.

Definition 3.1

In a *push* system, for each user in ascending order all input and output times are computed on each station in ascending order.

Then, a push strategy is based on two loops: the outer loop scans the sequence of users, the inner loop scans the sequence of stations.

According to this definition, in a push system each event is determined by the set of events already occurred in the system, and each time is set "as early as possible".

At the end of the procedure, once all times have been computed starting from the release dates, their feasibility with respect to the due dates is verified.

Definition 3.2

In a pull system, for each user in descending order all input and output times are computed on each station in descending order.

A pull strategy is based on two loops: the outer loop scans the inverse sequence of users, the inner loop scans the inverse sequence of stations.

In a pull system, event times are computed backward in time, and they are determined by the set of events that will occur forward in the future, on the basis of a "as late as possible" policy. At the end of the procedure, once all times have been computed starting from the due dates, their feasibility with respect to the release dates is verified.

While push and pull strategies are both valid solution methods for (SP), as proved in [12], their implementation in real service systems shows some structural differences, outlined by the following arguments.

First of all, the planning process, defined by (SP), can be seen as a global and off-line strategy based on a forecast of lead times, and does not take into account explicitly unpredictable problems (e.g. delays, stations downtimes) which can occur during the service. These local events ask for a (partial or global) replanning of the operations for all stages. In order to implement such a control function, the system must be managed employing dynamic strategies. In general, the introduction of dynamic management rules in a complex system is accomplished by decomposing the decision process into decision units, and representing how they interact, by means of physical and information flows.

Following the above statement, and according to definitions 3.1 and 3.2, in a *dynamic push rule*, each station serves a user, if any available, and pushes it downstream whenever it is possible, while in a *dynamic pull rule* each buffer calls the upstream station for a new user to be served whenever it has an empty location.

We are now able to show the differences between push and pull systems, analyzing the information flows network associated to their representation as network of processes.

In order to react to the undesired events mentioned above and to generate a new planning, in push systems the information flows network is characterized by multiple connections among service stages to carry the information on deadlines and ready times.

Moreover there is a centralized information and decision management center (Figure 3).

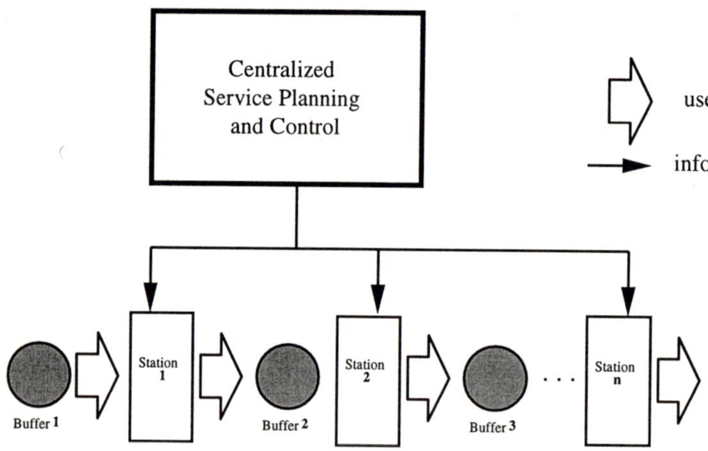

Figure 3. Information Flows in a Push System

On the contrary, in pull systems the information flows are completely embedded into the users flow, and, therefore, the only connections needed are between pairs of adjacent stages, from upstream to downstream. Moreover, no on-line centralized information and decision management is needed (Figure 4), and the system self-propagates the information needed to react to local unplanned problems, occurred in any point of the line.

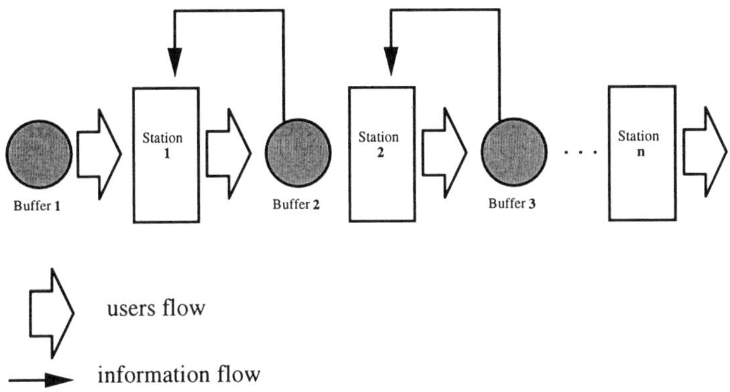

Figure 4. Information Flows in a Pull System

In the manufacturing systems area, a typical field in which pull-like strategies shows their superiority over the push-like systems is the control of work in process (WIP). In fact, while controlling the amount of inventory inside the system is a natural consequence of pull strategies, in push systems it requires a complex centralized replanning procedure involving heavy information exchange. In absence of this global intervention, in a push strategy each stage behaves without taking into account possible excess of inventory stored downstream, caused by unplanned perturbations. It is important to note that the last remark is confirmed by theoretical properties proved in [12], as far as minimization of WIP with perturbed processing times is concerned, and by computational results in extensive experiences in dynamic environments [2]. A smart implementation of pull strategies is represented by the celebrated Just In Time systems.

4. Flow Management Rules in Terminal Area

We first describe the current procedures employed in flow management on the ground of the statements of Section 3. It is quite clear that the arrival process can be properly modeled as a service network, whose service stations are the control facilities. Each aircraft is a user which has to cross a sequence of sectors, individuated by its route, each of them managed by one controller. At the end, the aircraft approaches the last facility which is the runway, and then it is cleared.

From now on we will refer to Figure 5, in which it is depicted the network of the main fixes of the FIR (Flight Information Region) of Rome, Fiumicino Airport. At each point of the network there is a holding pattern which acts as a buffer in which aircraft can wait for the next service, if the corresponding facility is busy.

Actually, the tactical planning and control phases in flow management are represented by slot assignment and aircraft scheduling in terminal area, respectively.

The first problem is generally formulated as an assignment problem [13], the second problem is a scheduling problem. The current procedures solve the two problems separately: slots are assigned by a centralized controller (Central Flow Management Unit of Eurocontrol in Europe), while the scheduling problem is solved in real time, on the ground of on-line measures, by the local controller in charge of landing [9]. These procedures perform a push-type strategy: each aircraft approaches the terminal control region, according to the arrival slot it has been assigned to by the off-line planning phase (Section 3), crosses a given sequence of fixes, and it is delayed by routing it on a holding pattern each time a central controller has to manage a congestion situation. The last operation is performed by a controller on the ground of on-line information on the available capacity and on the amount of traffic to manage. Moreover, even if controllers try to balance the workload of the fixes involved this is performed with an heavy information exchange, in a dynamic push fashion (Figure 3). Sometimes some informal flow management strategies, like speed reduction, are employed.

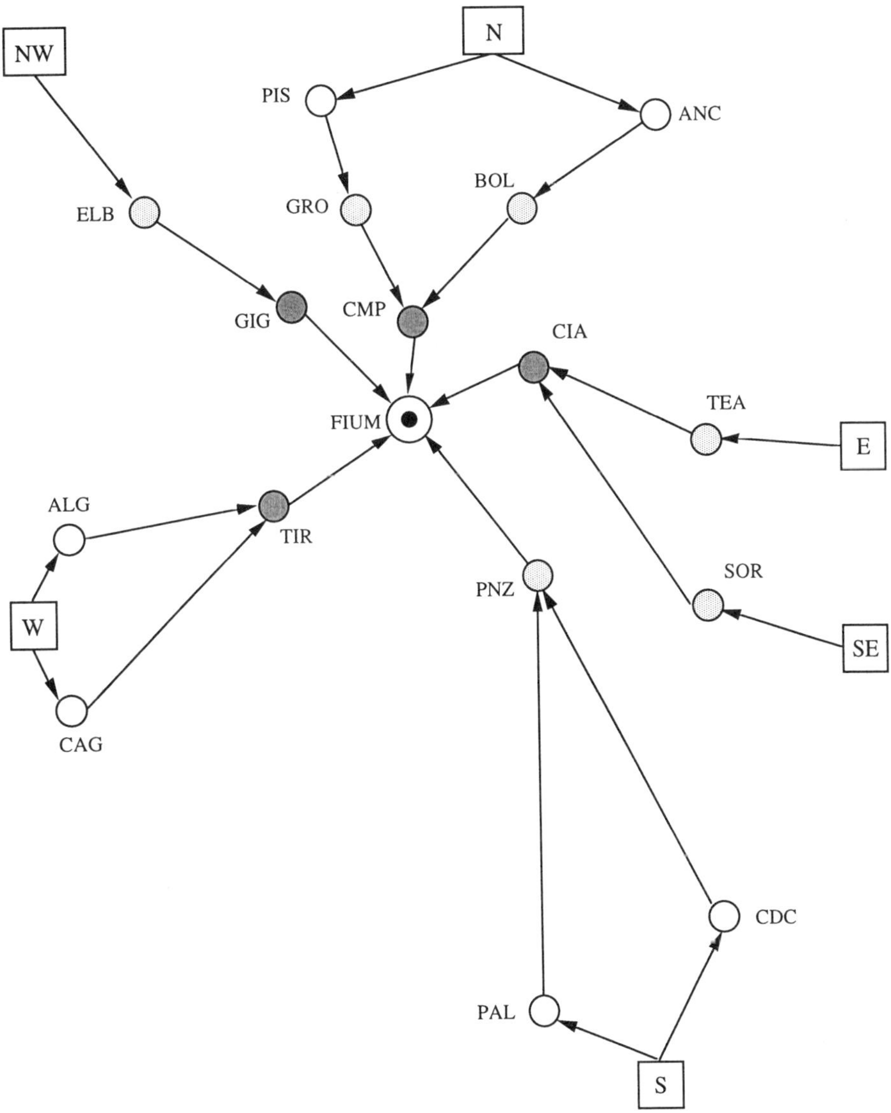

Figure 5. Simplified Layout of the FIR of Rome

Actually, it is not unusual to measure heavy airborne delays, which induce costs supported by airlines, and difficulties in maintaining safety standards.

The above arguments, and the statements of Section 3, allow us to introduce a pull rule in flow management, which will be compared, in Section 5, to the current push rule. Extensive experiments on real data will be carried out and analyzed.

A pull rule can be viewed as a simple and structured technique that allows an efficient control of the workload of the facilities, and of the number of aircraft kept in holding stacks (work in process), with a smart and decentralized information exchange. This strategy can be implemented with a systematic use of en-route speed reduction, and this can be considered a customization of the traditional pull rules employed in manufacturing, based on the fact that it is possible to increase the service time of any user at any service station.

We will now formally define the pull rule. Following the notation of Section 3 (Figure 4), from now on each holding pattern preceding a facility will be referred to as the corresponding buffer. For each station j, the pull rule can be detailed as follows.

PULL RULE:

INPUT	number of aircraft in the buffer j; max. number of aircraft Aj allowed (at any time) to wait in the buffer j; max. number of aircraft Bj allowed in order to not impose speed reduction;
DECISIONS	1. accept a new user in the buffer j; 2. impose a speed reduction;
CONDITIONS	1. number of aircraft currently in the buffer j less than Aj; 2. number of aircraft currently in the buffer j more than Bj ($Bj < Aj$);

DECISION TIME when station *j-1* is ready to start a new operation, when station *j* completes an operation;

ACTIONS station *j*-1 is allowed to process a user *i*; the service time of station *j-1* for the next user *i* is increased. After the end of the operation *o(i,j-1)* user *i* reaches the buffer *j*.

The speed reduction imposed is proportional to the number of users in the buffer. In the under exam case, we assumed that the flight time (time interval between the entering time in the FIR and the time the aircraft is cleared) can be increased of at most 15 min. in correspondence with the farthest entering point (S in Figure 5).

Defining the threshold values A_j, B_j for each buffer j is equivalent to controlling the workload, blocking or decreasing the upstream users flow.

This capability of maintaining the stability of users flow can be paid with the drawback, typical of pull-type rules, of increasing the completion time of all the operation (for a given set of users and given arrival times). In the next Section we show some experiments in which this kind of bad effects are negligible. On the contrary, the average lead time of users through the system decreases.

The comparison between the rules is performed on the ground of two indicators. The first considers the total amount of time spent by all the aircraft in holding patterns, which we want to minimize. The second one measures the stability of aircraft flow and the workload balance of the controllers.

5. Computational Results

The results are obtained simulating the FIR of Rome, whose simplified layout is reported in Figure 5. The experiments are based on the *flying-over* data provided us by ENAV (the national agency for flight support), relative to the 31 days of December 1995. For each flight originating, terminating, or crossing the Roman FIR, it is stored the time in which the aircraft enters and/or leaves the FIR. We filtered the data in order to consider only the arriving aircraft. Then, we selected 13 time windows of 3 hours, from 6:00 to 9:00 or from 14:00 to 17:00, characterized by the highest volume of traffic. In Table 1 simulation data are given: columns correspond to instances and rows correspond to time slots of 10 minutes; each entry represents the number of incoming aircraft; the last column reports the average number of incoming aircraft in each slot, while the last row reports the total number of incoming aircraft.

In Table 2 we show, with the same format, the geographical distribution of the traffic in the FIR: each row corresponds to an entry point in the FIR; each entry is the total number of aircraft entering the FIR from the corresponding point. The analysis of Table 2 shows that NN, NE, and NW entry points must accommodate a higher volume of traffic with respect to the other entry points.

The results confirm that the pull rule produces the expected effects in all the experiences.

First of all, the amount of time spent by all the aircraft in holding patterns decreases, as a consequence of the effective implementation of the speed reduction. This result is presented in Table 3 and Table 4, for push and pull rules respectively: for each instance (row) it is reported the time spent by all the aircraft in holding patterns and the time spent by all the aircraft flying inside the FIR on their route (not including the time in which aircraft are kept in a holding pattern), which we refer to as total flight time.

Notice that, in all the instances, the speed reduction policy increases the total flight time.

	1 dic 6 - 9	2 dic 6 - 9	4 dic 14 - 17	5 dic 6 - 9	6 dic 6 - 9	7 dic 6 - 9	12 dic 6 - 9
10	0	1	2	0	0	1	0
20	2	0	1	0	0	0	4
30	1	3	2	3	1	1	1
40	4	4	3	4	4	2	4
50	3	4	7	4	5	3	3
60	4	4	4	6	4	5	4
70	8	9	4	9	9	11	8
80	12	6	1	7	10	8	9
90	5	3	6	4	3	2	6
100	4	4	6	4	3	2	6
110	7	3	2	9	6	4	6
120	3	3	4	3	2	7	4
130	3	4	4	3	3	5	4
140	0	1	2	3	0	0	1
150	3	5	5	2	3	1	4
160	6	5	5	4	5	2	4
170	2	0	1	2	1	1	2
180	3	3	3	3	5	4	4
190	3	2	4	2	2	4	2
200	0	0	1	0	0	2	1
210	0	0	1	0	0	0	0
TOT.	73	64	68	72	66	66	75

	14 dic 6 - 9	21 dic 6 - 9	21 dic 14 - 17	22 dic 6 - 9	28 dic 6 - 9	29 dic 6 - 9	MED.
10	0	1	1	1	0	1	0.62
20	1	0	2	0	0	2	0.92
30	1	1	4	1	1	2	1.69
40	2	3	5	3	4	2	3.38
50	4	3	5	3	4	5	4.08
60	3	4	5	4	5	4	4.31
70	7	7	4	9	5	7	7.46
80	12	10	3	9	7	7	7.77
90	3	4	5	3	2	6	4.00
100	3	3	4	5	4	5	4.00
110	5	5	3	7	5	7	5.31
120	2	3	3	2	3	3	3.23
130	5	4	3	2	5	4	3.77
140	1	2	5	4	2	1	1.69
150	3	2	2	2	1	3	2.77
160	5	5	3	5	3	4	4.31
170	4	2	1	1	3	2	1.69
180	4	4	5	3	3	3	3.62
190	2	2	6	3	4	2	2.92
200	1	0	0	1	0	1	0.54
210	1	0	2	0	0	0	0.31
TOT.	69	65	71	68	61	71	68.38

Table 1. Traffic Data

	1 dic 6 - 9	2 dic 6 - 9	4 dic 14 - 17	5 dic 6 - 9	6 dic 6 - 9	7 dic 6 - 9	12 dic 6 - 9
NW	8	9	9	8	8	8	8
NN	22	17	27	24	22	18	25
NE	12	8	8	9	12	10	10
E	1	4	5	4	1	5	4
SE	7	6	4	8	6	6	8
S1	10	6	5	7	4	7	8
S2	3	3	2	3	2	3	3
SW	4	5	3	4	4	3	4
W	6	6	5	5	7	6	5

	14 dic 6 - 9	21 dic 6 - 9	21 dic 14 - 17	22 dic 6 - 9	28 dic 6 - 9	29 dic 6 - 9	MED.
NW	9	6	10	6	7	7	7.92
NN	19	21	31	19	17	19	21.62
NE	10	9	7	12	8	10	9.62
E	5	5	3	2	4	3	3.54
SE	7	7	5	6	6	6	6.31
S1	6	6	5	9	7	11	7.00
S2	3	3	2	3	3	4	2.85
SW	3	3	4	5	3	5	3.85
W	7	5	4	6	6	6	5.69

Table 2. Geographical Distribution of the Traffic

INSTANCE	TOTAL TIME IN HOLDING (min.)	TOTAL FLIGHT TIME (min.)	COMPLETION TIME (min.)
1 dic 6 - 9	1570	1581	229
2 dic 6 - 9	672	1370	207
4 dic 14 - 17	305	1441	226
5 dic 6 - 9	1465	1538	222
6 dic 6 - 9	1036	1419	212
7 dic 6 - 9	1000	1404	217
12 dic 6 - 9	1364	1604	227
14 dic 6 - 9	1468	992	225
21 dic 6 - 9	824	1385	212
21 dic 14 - 17	368	1510	233
22 dic 6 - 9	1223	1468	222
28 dic 6 - 9	273	1307	212
29 dic 6 - 9	1194	1538	222

Table 3. Holding Times and Flight Times for Push Rule

INSTANCE	TOTAL TIME IN HOLDING (min)	TOTAL FLIGHT TIME (min)	COMPLETION TIME (min)
1 dic 6 - 9	1314	1992	231
2 dic 6 - 9	428	1628	208
4 dic 14 - 17	205	1645	226
5 dic 6 - 9	1195	1935	226
6 dic 6 - 9	740	1773	214
7 dic 6 - 9	753	1712	219
12 dic 6 - 9	1122	1970	229
14 dic 6 - 9	701	1793	227
21 dic 6 - 9	622	1700	214
21 dic 14 - 17	226	1757	233
22 dic 6 - 9	927	1824	223
28 dic 6 - 9	185	1480	213
29 dic 6 - 9	910	1855	222

Table 4. Holding Times and Flight Times for Pull Rule

Another desirable result of pull strategy is the stability of flow, which induces a balanced workload for all the control facilities. This is pointed out by Table 5 and Table 6 in which we report the average number of aircraft in the sequence of fixes PIS - GRO - CMP - FIU (Figure 5), for push and pull case respectively. Notice that the pull rule produces a more balanced distribution of traffic, spreading the aircraft among all the fixes defining the route and controlling traffic peaks.

INSTANCE	PIS	GRO	CMP	FIU
1 dic 6 - 9	0	0	3.17	2.59
2 dic 6 - 9	0.03	0	0.65	1.95
4 dic 14 - 17	0.08	0	0.05	1.12
5 dic 6 - 9	0.02	0	3.28	2.36
6 dic 6 - 9	0	0	2.29	2.08
7 dic 6 - 9	0.04	0	1.76	2.14
12 dic 6 - 9	0.03	0	2.96	2.45
14 dic 6 - 9	0.02	0	1.63	2.20
21 dic 6 - 9	0.01	0	1.55	2.01
21 dic 14 - 17	0.07	0	0.16	1.23
22 dic 6 - 9	0.04	0	2.47	2.34
28 dic 6 - 9	0	0	0.08	1.08
29 dic 6 - 9	0	0	1.92	2.55

Table 5. Average Number of Aircraft in Some Fixes for the Push Rule

INSTANCE	PIS	GRO	CMP	FIU
1 dic 6 - 9	0.29	1.16	1.83	1.55
2 dic 6 - 9	0.03	0.06	0.59	1.28
4 dic 14 - 17	0.08	0.03	0.12	0.68
5 dic 6 - 9	0.43	1.43	1.59	1.35
6 dic 6 - 9	0.03	0.68	1.22	1.29
7 dic 6 - 9	0.05	0.26	1.25	1.45
12 dic 6 - 9	0.29	0.98	1.59	1.37
14 dic 6 - 9	0.02	0.22	1.18	1.33
21 dic 6 - 9	0.01	0.14	1.01	1.42
21 dic 14 - 17	0.07	0.01	0.17	0.70
22 dic 6 - 9	0.04	0.39	1.66	1.34
28 dic 6 - 9	0	0	0.14	0.71
29 dic 6 - 9	0	0.32	1.45	1.46

Table 6. Average Number of Aircraft in Some Fixes for Pull Rule

Moreover, we confirm that the typical drawback of the pull rules (increasing the completion time of all the operations), did not arise, due to the effective customization of the rule. Completion time did never overcome 233 min, with negligible differences between push and pull systems (Table 3-4).

We conclude this Section with few details on the implementation of the simulation model. The model is implemented with the discrete event simulation tool WITNESS 6.0, running on a IBM RISC 6000 3AT workstation. Average running time of our experiments is around 0.5 min.

6. Conclusions

In Section 4 we outlined the gap between slot assignment procedures and local management inside the terminal control region, mainly due to the different decision environment. Although it could be difficult to realize a global strategy, the introduction of the pull rule allows to take into account the two phases in an integrated framework.

Our results show that the implementation of the pull strategy can be very effective. Moreover, in order to properly evaluate them, it must be taken into account that the pull rule has been compared with a pure push rule without introducing any informal procedure [9] that in practice can be employed by a controller for congestion handling. In other words, the local controller sometimes behaves in a pull fashion. Our approach can be viewed as a framework which co-ordinates the local decisions of the different controllers.

Acknowledgements

A particular acknowledgement to Massimo Riccio for his contribute in the implementation of the simulation model.

References

[1] A. Agnetis, M. Lucertini, F. Nicolò, Flow management in Flexible Manufacturing Cells with Pipeline operations, *Management Science*, vol.39, n.3, pp. 294-306, march 1993.

[2] A. Agnetis, M. Lucertini, S. Nicoletti, F. Nicolò, G. Oriolo, D. Pacciarelli, A. Pacifici, E. Pesaro, F. Rossi: *Material flow management in an automobile assembly plant*. To appear.

[3] A. Agnetis, M. Lucertini, *Design criteria for flexible production systems based on non-simultaneous demand models*, Proceedings of the 2nd International Conference on CIM, Troy, NY, May 1990.

[4] G. Andreatta, G. Romanin Jacur, Aircraft flow management under congestion, *Transportation Science*, vol.21, n.4, pp.249-253, 1987.

[5] L. Bianco, A. Odoni (Eds.), *Large scale computation and information processing in air traffic control*, Springer-Verlag, 1993.

[6] L. Bianco, A. Odoni, G. Szego, *Flow control of congested areas*, Springer, 1987.

[7] R.B. Chase, N.J. Aquilano, *Production and operations management*, Irwin, 1989.

[8] M. Carnevale, M. Lucertini, S. Nicosia (Eds.), *Modelling the innovation*, North-Holland, 1990.

[9] G.A. Hocker, *Airport Demand and Capacity Modeling For Flow Management Analysis*, Master Thesis, MIT, Cambridge, Massachusetts, 1994.

[10] A. Kumar, P.S. Ow, M.J. Prietula, Organizational simulation and information systems design: an operations level example. *Management Science*, 39, 2, pp.218-240, 1993.

[11] A.J. Kulikowski, *Traffic flow management in today's ATC system*, J. of ATC, Jan.-March 1991.

[12] M. Lucertini, F. Nicolò, W. Ukovic, A. Villa, Models for Flow Management, to appear on *International Journal of Operations and Quantitative Management.*

[13] P.B. Vranas, D.J. Bertsimas, A.R. Odoni, The multi-airport ground-holding problem in air traffic control, *Operations Res.*, vol.42, n.2, pp.249-261, 1994.

THE CENTER-TRACON AUTOMATION SYSTEM: SIMULATION AND FIELD TESTING

Dallas G. Denery

Chief, Air Traffic Management Branch,
Ames Research Center, Moffett Field, CA 94035-1000, USA

Heinz Erzberger

Senior Scientist, Air Traffic Management Branch,
Ames Research Center, Moffett Field, CA 94035-1000, USA

A new concept for air traffic management in the terminal area, implemented as the Center-TRACON Automation System, has been under development at NASA Ames in a cooperative program with the FAA since 1991. The development has been strongly influenced by concurrent simulation and field site evaluations. The role of simulation and field activities in the development process will be discussed. Results of recent simulation and field tests will be presented.

1. Introduction

A system for the automated management and control of terminal area traffic to improve productivity, referred to as the Center-TRACON Automation System (CTAS), is being developed at NASA Ames Research Center under a joint program with the FAA [1]. CTAS consists of three types of integrated tools that provide computer-generated advisories for both en-route and terminal area controllers to manage and control arrival traffic efficiently. The first tool, the Traffic Management Advisor (TMA), generates runway assignments, landing sequences, and landing times for all arriving aircraft, including those originating from nearby feeder airports [2]. TMA also assists in runway configuration control and flow management. The second tool, the Descent Advisor (DA), generates clearances for the en-route controllers handling arrival flows to metering gates [3]. The DA's clearances ensure fuel-efficient and conflict free descents to the metering gates at specified crossing times. The third tool, the Final Approach Spacing Tool (FAST) provides terminal area controllers with heading and speed advisories to help produce an accurately spaced flow of aircraft onto the final approach course [4].

The underlying premise behind the design of CTAS has been that successful planning of traffic in capacity constrained airspace requires the ability to accurately predict future traffic situations. The technology for accurate prediction of trajectories was developed in the early 1970s and has been incorporated in modern flight management systems. Data bases consisting of several hundred aircraft performance models, airline preferred operational procedures and a three dimensional wind model support the trajectory prediction capabilities within CTAS. (This is discussed in [7]).

The primary research effort within CTAS has been the design of a set of automation tools that make use of this trajectory prediction capability to assist the controller in overall management of traffic. The two criteria upon which success is judged are controller acceptance and improvement in traffic flow as measured by reduced delays and improved aircraft operating efficiencies. Because of the complexity of the air space system, the approach taken has been to adopt a “design a little, test a lot” philosophy with real-time simulation and field testing included as an integral part of the design process. Analysis of real-time data and fast-time simulation methods are used to extrapolate the results of the field tests.

The purpose of this paper is to review the process used in the development of CTAS and provide examples of the role of real-time simulation, field testing, and fast-time simulation. The paper will first discuss the overall technical approach. To illustrate the approach, the FAST development will be reviewed. The DA tool is somewhat different from FAST in that it allows more strategic control. This has led to some differences in the DA development approach that will be discussed.

2. Technical Approach

The overall technical approach is shown in figure 1. Instead of following the more traditional sequential-approach, the requirements, design, simulation, and operational tests are conducted concurrently with a high level of interaction. Analysis of real-time simulation and live traffic data are used with fast-time simulation to quantify and extrapolate the performance of the system. A primary advantage of this approach is the involvement of controllers and pilots throughout the development.

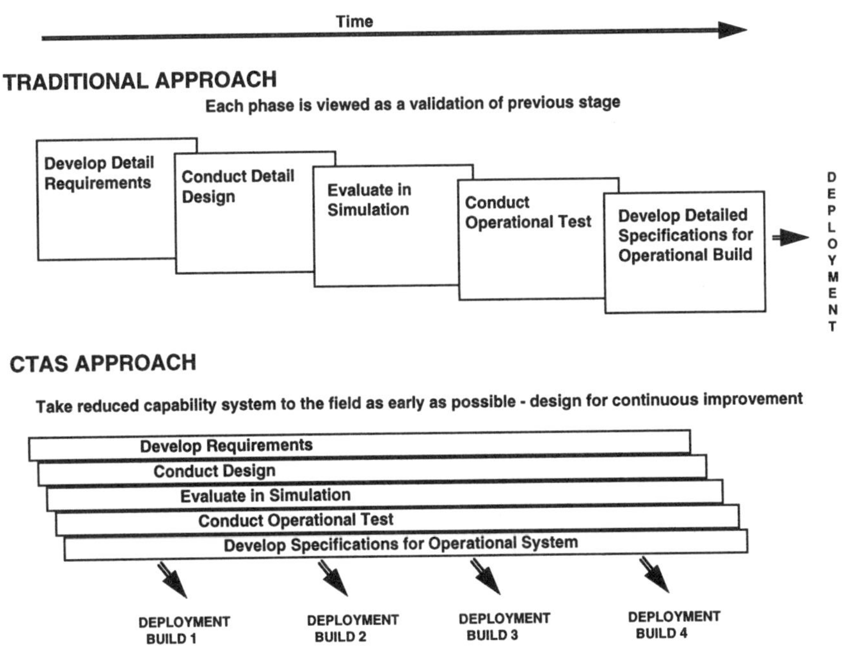

Figure 1. Programmatic approach

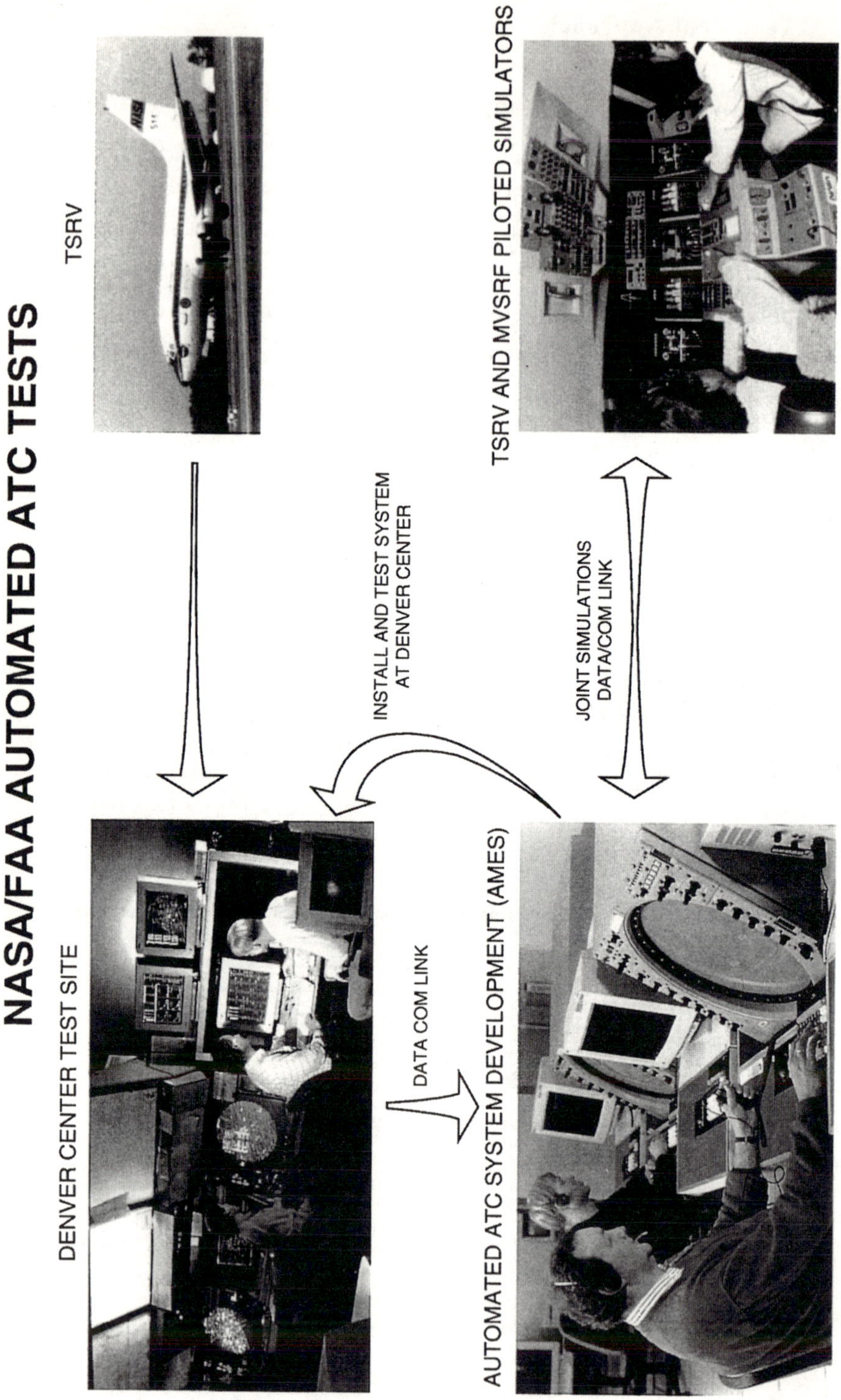

Figure 2. Research facility

The research facility established to support this approach is illustrated in figure 2. The primary ATC simulation was developed at Ames. It includes an air traffic simulation using pseudo-pilots and an ATC facility simulation. Both are hosted on a network of workstations. To study controller display integration issues, two terminal area radar displays (Fully Digital ARTS Display System, FDADS) are integrated into the network. To investigate specific air-ground communication and traffic management issues, links were established with existing full piloted simulators located at the Ames and Langley Research Centers. To understand actual traffic situations and to support shadowing evaluations, live radar connections were established, first with Denver Center and then expanded to include the Fort Worth Center and the Dallas/Fort Worth terminal area (TRACON). To understand weather and evaluate its effect on the trajectory prediction capability of CTAS, connections were established to receive weather information for both the Denver and Dallas/Fort Worth areas. We are currently receiving "rapid update cycle" weather data. Field tests are under way at Denver and Dallas.

3. Development of FAST

The steps taken in the FAST development are illustrated in figure 3. Fast-time simulation, real-time simulation, and live traffic testing in shadow-mode have been used throughout the development [5]. Operational testing has been maintained as a target but has been delayed until the system design issues identified in simulation and shadow-mode testing are resolved. Controllers have been involved throughout the process. Initial studies considered a generic airspace designed to evaluate basic concepts. As the program progressed, the effort addressed more realistic environments based on the Denver and Dallas/Fort Worth areas.

3.1 *FAST Description*

FAST is a tool for aiding the terminal area controller in setting up the optimal landing sequence, selecting the most appropriate runway and providing the controller with turn and speed advisories to produce an accurately spaced flow of aircraft onto the final approach course [4]. The sequence and runway advisors are referred to as "passive FAST." The turn and speed advisories are referred to as "active FAST." Both passive and active FAST advisories are based on trajectories that have been computed to be conflict free for the duration of the flight path. These trajectories and advisories are continually updated based on new radar track data (every 4.7 seconds) and on inference of controller intent. More details on FAST are contained in [4] and [6]. The trajectory prediction computations are reviewed in [7].

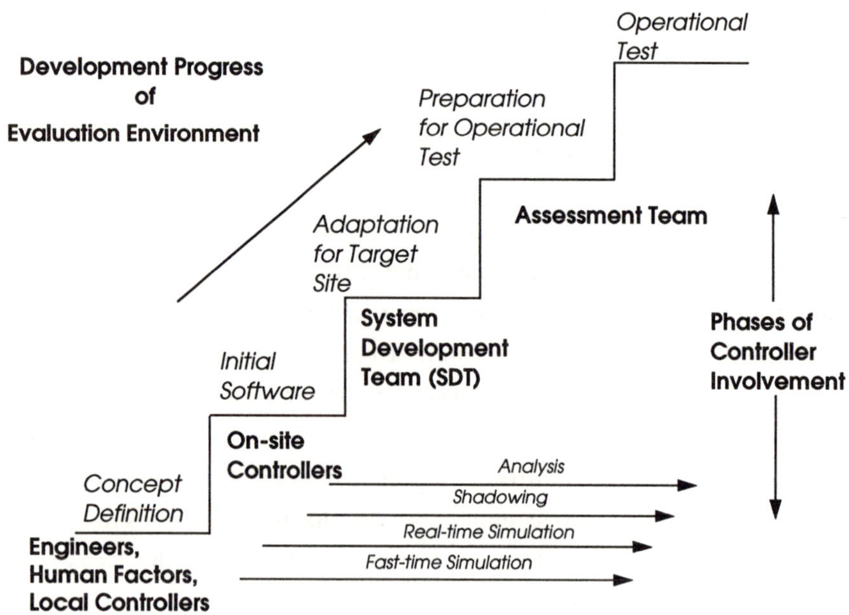

Figure 3. FAST development process

As an example of the development process, we will review the developments of the sequence and scheduling logic and the runway allocation logic.

3.2 Sequencing and Scheduling Algorithm

The sequencing and scheduling problem addressed within FAST are illustrated in figure 4. In the initial design, the sequence and schedule were optimized to assure minimum delays based on separations at the threshold. The speed and turn advisories were computed to assure efficient and conflict free flight [8]. To achieve minimum delays, the system would allow overtakes upstream in the traffic flow. As the simulation was adapted to be more representative of Denver and Dallas/Fort Worth, it became apparent that additional sequence constraints would be required to allow the controller to maintain a coherent view of the traffic situation. This led to the development of a trajectory segment based ordering logic that under certain conditions would maintain sequences established prior to merging on final [4,6]. The segment based ordering method allows the overtake of one aircraft by another if there is a sufficient reduction in delay but it restricts the conditions under which this reordering may occur. The logic for the reordering was derived from over 2000 hours of real-time simulations involving controllers from Dallas/Fort Worth. It is imbedded in the CTAS code in the form of fuzzy logic. An example of the resulting logic for a reordering is shown in figure 5. Without going into the details, the logic for determining whether to allow an overtake depends on the relative position of two aircraft scheduled for the same segment in the TRACON (i.e., downwind, final, etc.), their speed differences, and the potential delay savings. If the trailing aircraft falls above the curve in figure 5, it is rescheduled. Subsequent analysis and fast-time simulation have shown that these additional constraints impose a negligible penalty on overall performance.

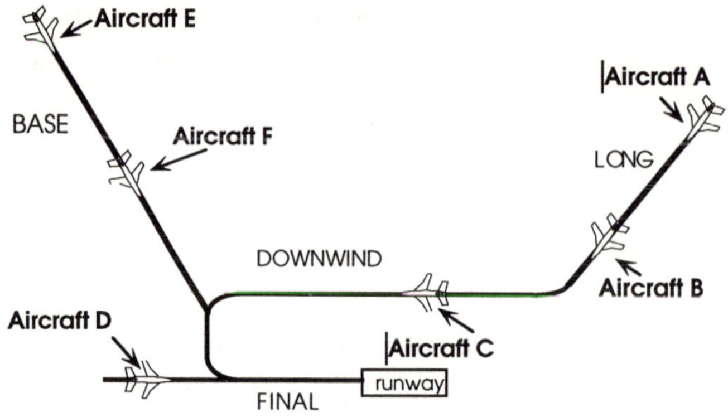

Figure 4. FAST sequencing and scheduling

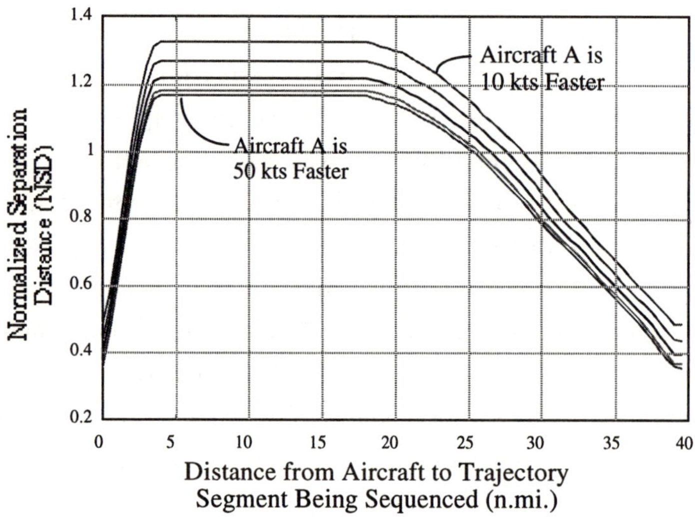

Figure 5. Knowledge based sequencing decision curves

3.3 Runway Allocation

The runway allocation algorithm has evolved from an initial algorithm that was designed to optimize a single functional [9], to an algorithm that is more consistent with current procedures, provides improved controller awareness, and allows consideration of multiple performance metrics [4,6]. The current method begins with a nominal runway assignment based on published procedures at the particular airport. A decision tree is entered which branches through alternative runways, entry gate to the TRACON, aircraft type, and finally ends with a minimum global delay reduction required for a runway change. The overall benefit due to a runway change is computed and compared with the predetermined minimum delay reduction. If the delay reduction exceeds the minimum delay, the change is made.

Real-time simulation and shadow mode operation have demonstrated the value of the runway allocation algorithm in two areas. The most significant improvement has been in elevating the performance of all controller crews to that of the best controller crews. Based on real-time analysis, to be discussed later, there is a large variation in the utilization of multiple runways as a function of different controller teams. A second area of improvement, even for the better controller teams, has been the identification of runway changes based on traffic information not available to the specific sector controller. This is illustrated in figure 6. The arrival sector controller may not be aware of the additional traffic coming in on the upper right side and as a result assign the aircraft on the lower right to the left runway. Due to a more global awareness of traffic, FAST would be able to determine an advantage in switching the aircraft to the right runway.

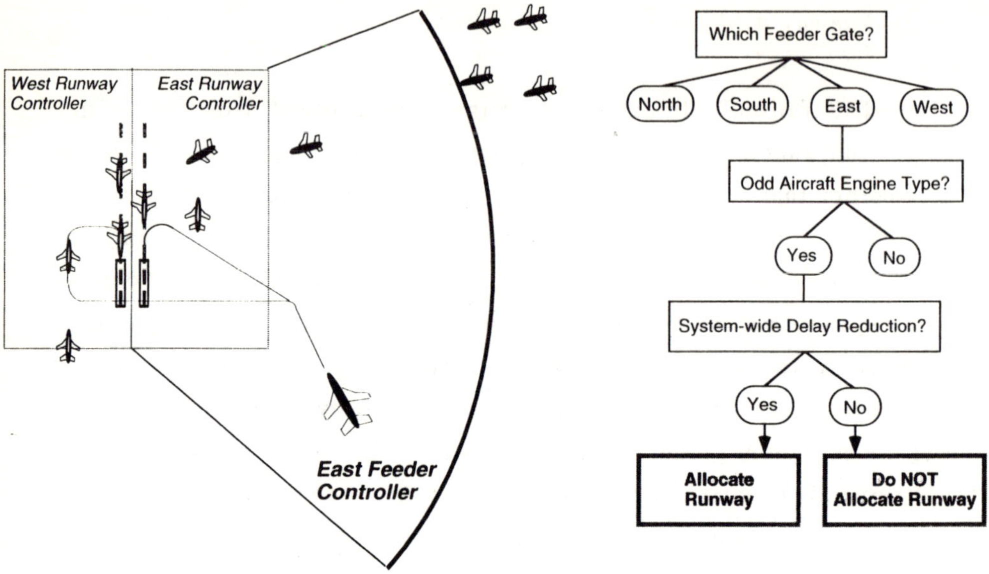

Figure 6. Knowledge based runway allocation

3.4 Human Factors Assessments

The CTAS development has incorporated the expertise of the end-user from the very beginning. The design has been guided by the premise that automation should extend a controller's ability to manage traffic rather than change a controller's overall responsibilities.

To maintain this focus, a human factors team was assembled by the FAA Liaison Office at Ames to work directly with the engineering staff. Additionally, the FAA established a team of experienced and highly skilled controllers to work with the team. Controller acceptance has been evaluated through observation, conducting interviews, recording the number of communications, and taking controller evaluations using various rating scales. In an attempt to standardize controller ratings a "Controller Acceptance Rating Scale (CARS)," is being developed [5]. An early version of the CARS is shown in figure 7. The idea is borrowed from the "Cooper Harper Rating" that has been very successful in standardizing pilot ratings for aircraft handling qualities [10].

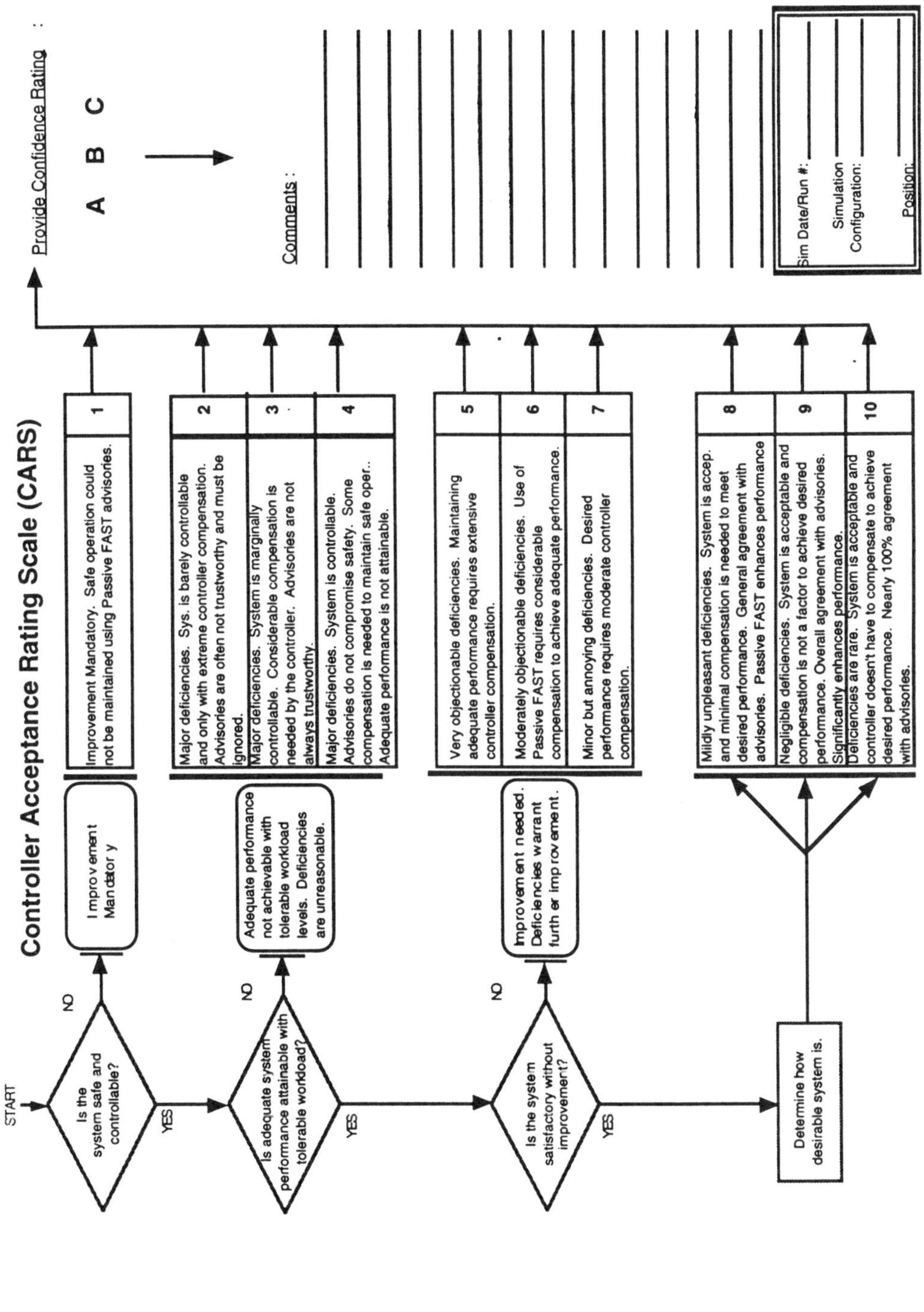

Figure 7. Controller acceptance rating

3.5 Analysis of Real-Time Data

So far, we have been discussing the development process. To understand whether the concept will provide benefit, techniques for analyzing real-time data are required to assure that the system will perform as expected in the real-world and to assist in quantifying potential benefits through use with fast-time simulation.

The real-time analysis conducted in support of CTAS is to be published this fall in an article by M. Ballin and H. Erzberger [11]. Two examples of this analysis are included here. First is the method used to calculate the arrival time errors at the feeder-fix into the terminal area. Based on fast-time simulation, Erzberger and Neuman have shown that the magnitude of these errors directly affect the portion of total delay that should be absorbed in the terminal area or TRACON [12,13]. The second is the method used to measure inter-arrival spacing at the threshold for different aircraft combinations, i.e. heavy followed by heavy, large followed by small, etc. These data are necessary to understand the delay reduction potential of improved sequencing and spacing and runway assignment.

Figure 8 shows a composite plot of flights into DFW taken over a 140 minute interval involving a major rush. A program has been developed to assist developers in analyzing these data [14]. The analysis program is constructed so that the CTAS estimated time of arrival (ETA) at the feeder fix, computed at any point along the trajectory, can be compared with the actual crossing time. The program is further refined so that a researcher can call up a specific trajectory to identify possible causes of any major error in the ETA. This tool has been invaluable in improving the overall robustness of the trajectory prediction algorithms.

An example of the use of this tool for obtaining statistical data on ETA errors is shown in figure 9. It should be noted that the curve appears to be the superposition of two error sources, one with a Gaussian distribution and one with a Poisson distribution. If the Gaussian portion is attributed to errors in the ETA calculations where the flight is not affected by controller-induced delays and the Poisson portion is attributed to delays inserted to coordinate traffic flow, we can make a first order estimate of ETA accuracy achievable with an effective traffic management tool.

Figure 10 shows a composite plot of flights into the terminal area. Here it is much more difficult to automatically sort through the data to achieve meaningful statistical results regarding ETA's at the threshold or estimates of the inter-arrival spacing. The tool must ignore all aircraft that are not landing, and it must identify the most likely runway for each landing aircraft. The greater the number of mistakes, the less valid the analysis.

Shown in figure 11 is an example histogram of inter-arrival spacing for aircraft having a legal separation of 2.5 n. mi. The few cases where separations were less than 2.5 n. mi. do not imply violations. Under current rules, as soon as the pilot has the runway in view, the pilot can declare VFR. Again, the curve seems to be a superposition of a Gaussian and Poisson distribution. In this case, it is assumed that the Gaussian portion represents the controller precision in spacing aircraft onto the final approach path given a steady stream of traffic and the Poisson portion represents those pairs where there were natural gaps. From these data, we can infer the controller target point, the errors that can be expected about the target point, and the buffer that can be used to model the controller's behavior. The potential for improvement is computed based on the expectation of achieving a reduced variance and buffer through the use of FAST advisories and the elimination of unnecessary gaps by improved runway balancing and delivery of aircraft into the terminal area.

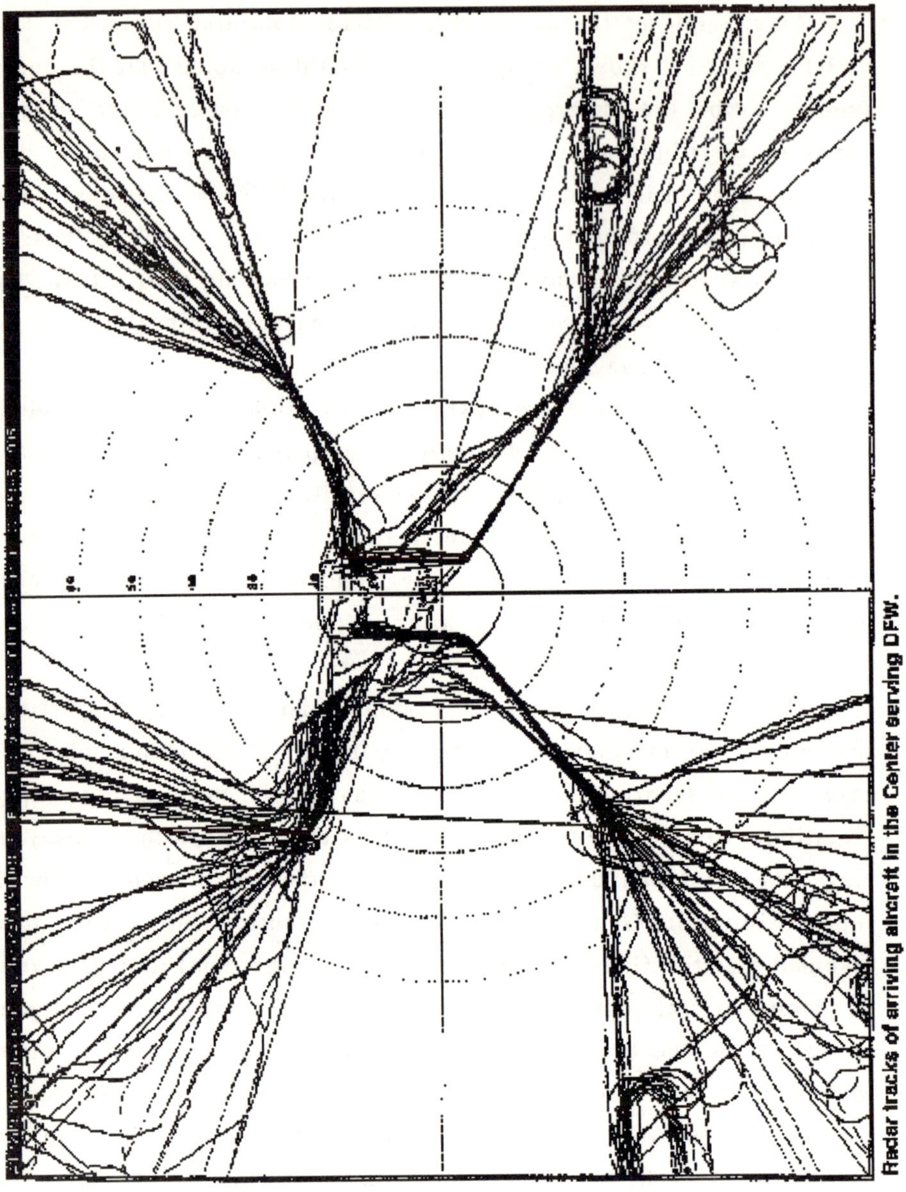

Figure 8. Composite of Center flights feeding DFW

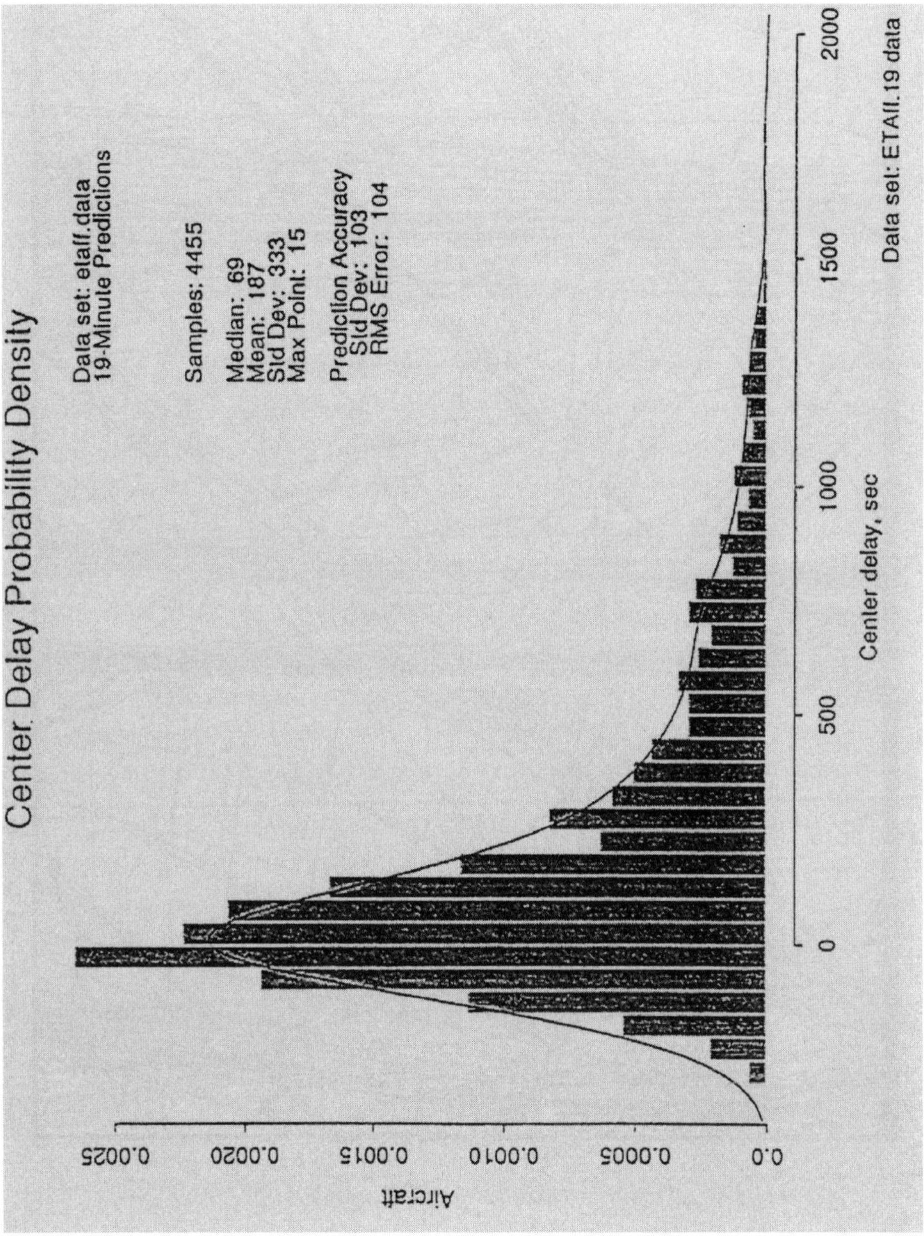

Figure 9. Histogram of ETA errors at meter fix

Figure 10. Composite of arrival flights.

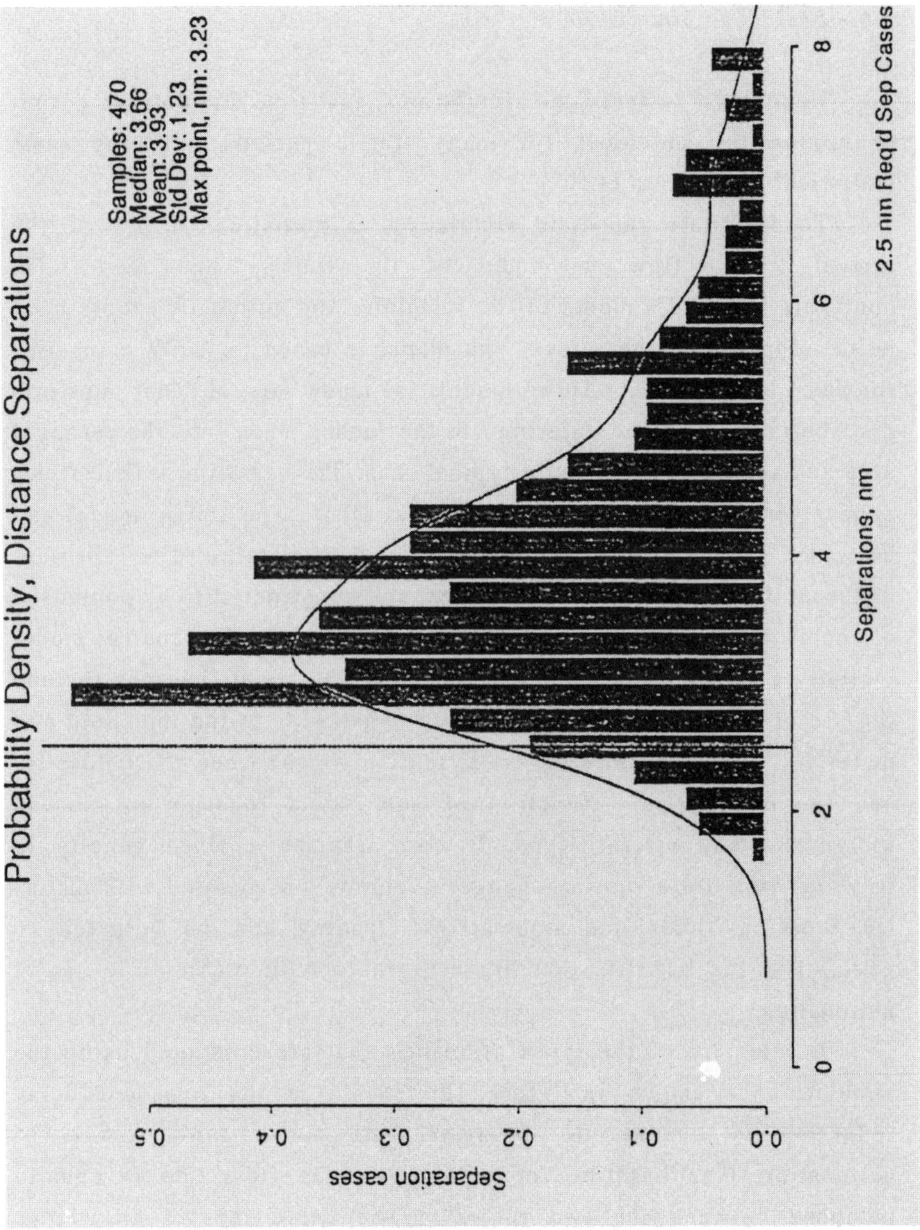

Figure 11. Histogram of interarrival spacing for aircraft pairs requiring a 2.5 n. mi. separation

3.6 Fast-Time Simulation

In contrast to real-time simulations, fast-time simulations permit examining the outcomes for many traffic periods with the same statistical parameters [12].

To facilitate fast-time simulation, a statistical model of the arrival traffic flow, a model of the runway and feeder fix configuration, and a model of the scheduler and automation tools must be developed. For these tests, the model is based on DFW using two runways. The traffic flow model is made up of four uniform distributions of traffic entering via the feeder fixes into the terminal area and scaled to represent a typical rush. The resulting traffic flows appear very similar to those observed at DFW. The traffic model can be scaled to represent different levels of traffic, tailored to represent different densities at individual gates, and constructed to be composed of specified percentages of aircraft types. The air traffic control model includes a set of simplifying assumptions. The simplifications include the use of fixed time based separation constraints at the threshold and meter fix, constant times for an aircraft to fly between the feeder-fix and the runway as a function of gate, as a function of runway assignment and aircraft type. It also assumes a fixed penalty in traversal time for a runway change. Accuracy's associated with meter-fix crossing times and inter-arrival spacing can be adjusted to understand the benefits that are achievable with different levels of automation.

A summary of the types of results that are computed using the simulation is shown in figure 12. Shown is the expected delay reduction as a function of arrival rate for different levels of automation. The baseline represents a traffic flow that is equally balanced between the two runways. The curve labeled knowledge based runway allocation, KBRA, shows the improvement achievable by allowing switches to the runway assignment to even out irregularities in the prearranged flow to the two runways. Similarly, the curve labeled "Active FAST" shows the further improvement due to more precise control of spacing on final.

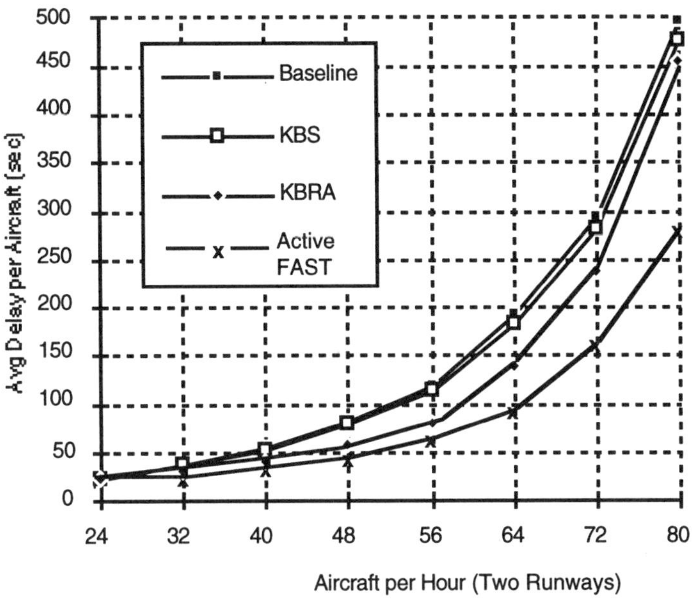

Figure 12. Delay reduction potential of CTAS

In recent studies, Erzberger and Neuman have used fast-time simulation to study the effect of errors in the meter-fix crossing time on (1) total delay and (2) the allocation of total delay between the Center and the terminal area (TRACON). The basic idea is that in the absence of uncertainty in the meter fix crossing time, none of the delay should be taken in the TRACON due to increased fuel burn rate at low altitudes. However, as errors are introduced into the meter-fix crossing time, if some delay is not allocated to the TRACON there may be a missed landing opportunity or at least an unnecessarily large gap in the arrivals. This will result in larger total delays and increased total cost. This study will be presented as a subject of an AGARD lecture series to be presented by Dr. Erzberger later this year [13].

4. Development of DA

Although the DA development has been very similar to that taken in FAST, there are some fundamental differences. The major difference is that DA attempts to develop strategic clearances requiring few changes during the descent. This places a more stringent requirement on the trajectory prediction accuracy and has led to (1) the inclusion of pilots as well as controllers throughout the development process and (2) the conduction of limited field evaluations at an FAA facility during the early phases of development to validate procedures and trajectory prediction accuracy.

4.1 DA Description

The Descent Advisor is a set of automation tools to assist the controller in delivering aircraft to the meter fix at a specified time and with specified crossing restrictions in a manner consistent with preferences of the aircraft operator. The advisories are computed to be consistent with the specific aircraft performance and on-board equipment (flight management system, FMS, or non-FMS) and computed to be conflict free for the duration of the trajectory. The advisories are refreshed based on continuous analysis of new radar data and detection of non-conformance to clearances. The advisories include cruise Mach number, descent speed profile, top of descent for non-FMS equipped aircraft, path stretching and route off-set, and direct-to heading advisors for non-FMS equipped aircraft. To illustrate the difference in approach between DA and FAST, we will review the recent field test conducted at the Denver Center.

4.2 DA Field Test

The objectives of the field test were to evaluate the ability of CTAS to accurately predict the trajectories resulting from DA advisories, to evaluate the benefits derivable from on-board FMS capabilities, and to develop compatible air/ground procedures [15].

The test performed in September 1994, involved 97 United Air Lines flights into Denver and 26 runs using the Langley Research Center's Terminal Systems Research Vehicle (TSRV) aircraft. The United flights were included to test the robustness of the system to different aircraft types, different wind conditions, different crews, and different levels of flight management equipment. The TSRV was included to provide detailed information on the winds, and to assess the accuracy and sources of errors in the trajectory prediction algorithms. The tests were conducted with airspeeds varying between 240 and 320 KIAS. Participating United Airlines flights included B757 and B737 aircraft equipped with flight management systems and B727 and B737 without flight management systems. The TSRV was flown as a conventionally equipped aircraft and an FMS equipped aircraft.

The test was configured to negate the impact on air traffic or air carrier operations. The configuration is shown in figure 13. A DA test station was set up in the Traffic Management Unit of the Denver en-route center. The existing CTAS system that supports TMA at Denver was used. The DA advisories were transmitted to a test engineer located at the sector controller position. The test engineer passed the advisory to the sector controller in a written script. The sector controller then issued the advisory to the participating flight.

An example of a DA advisory for an unequipped aircraft would be:

> "UAL 123, begin descent 70 miles from the Meeker VORTAC; descend at 280 knots; if unable advise."

An example DA advisory for a FMS equipped aircraft would be:

> "UAL 123, descend at pilot's discretion, descend at 280 knots; if unable advise."

The exact phraseology and procedures were carefully coordinated between the facility and United Airlines.

Examples of the data collected are shown in figure 14. Both horizontal and vertical profile data as well as ETA errors were recorded. The data shown are for an aircraft with an FMS and for an aircraft without an FMS. A summary of the accuracy achieved at the meter fix is shown in table 1 in the form of mean and root mean square (rms). In all cases the CTAS prediction was within 20 seconds.

The FMS in the TSRV predicted crossing time is also shown for comparison.

As previously noted, fast time analysis has indicated a strong relation between operational benefits and the accuracy with which aircraft are delivered across the meter-fix. Based on a preliminary extrapolation of this analysis, the better than 20 second delivery accuracy shown above to be achievable with DA, together with the benefits derivable with FAST and TMA are estimated to be in the order of $33M per year at the DFW airport. These data are being used by the FAA to develop a comprehensive assessment of the benefits achievable with CTAS.

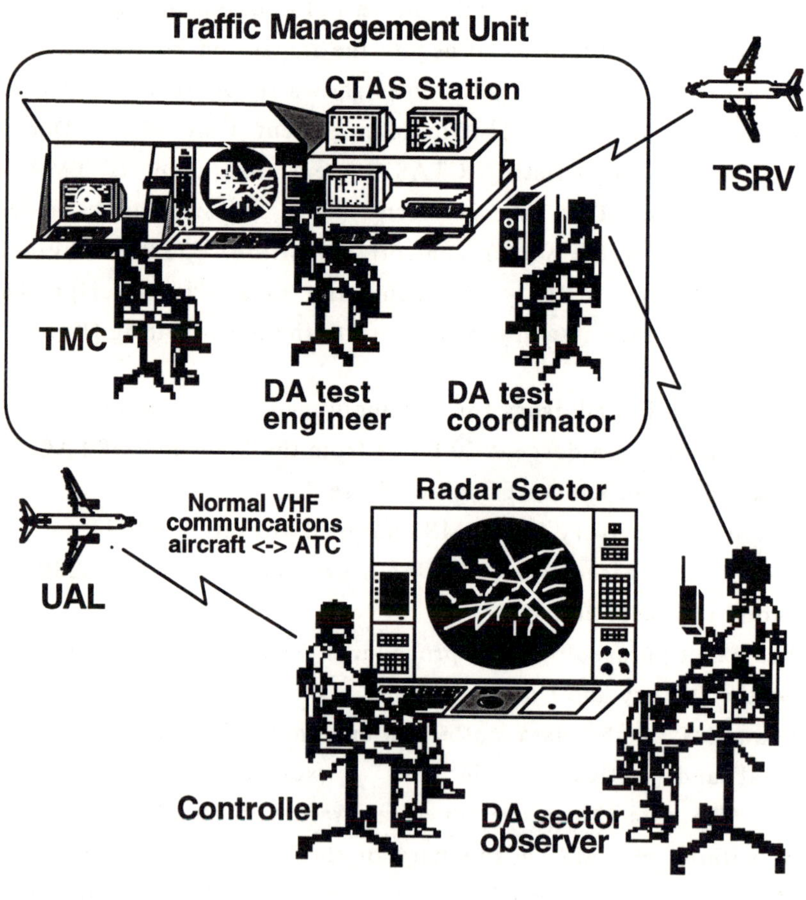

Figure 13. DA test configuration

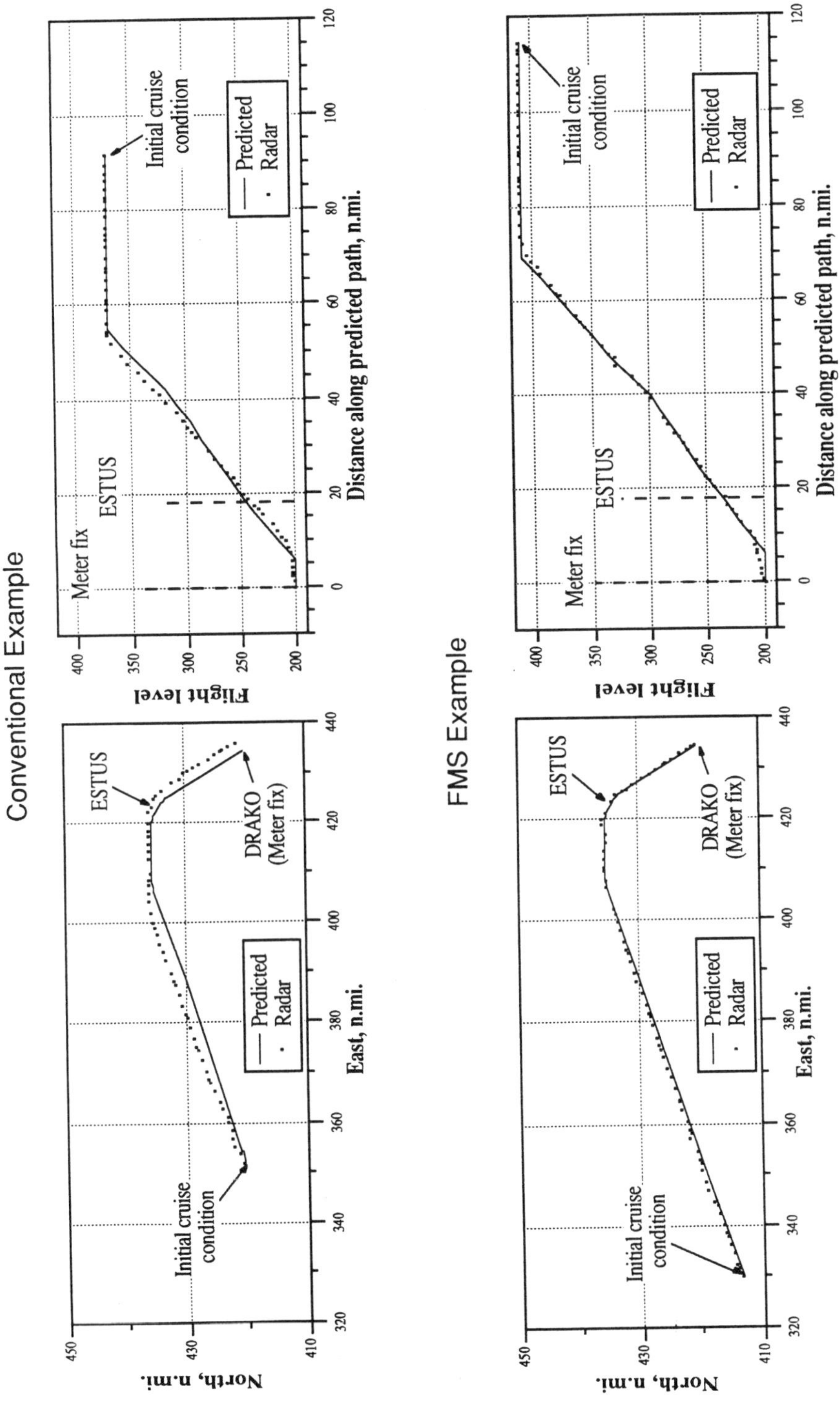

Figure 14. Trajectory data collection

	TSRV aircraft		UAL aircraft
Guidance mode	FMS prediction	CTAS prediction	CTAS prediction
All	8.8 mean, 10.5 rms	–2.3 mean, 12.5 rms	2.4 mean, 13.1 rms
non-FMS	16.8 mean, 9.4 rms	1.7 mean, 10.0 rms	7.4 mean, 14.3 rms
FMS	4.9 mean, 9.4 rms	–6.3 mean, 12.4 rms	–2.5 mean, 10.0 rms

Table 1. Meter fix crossing time accuracy (seconds)

5. Concluding Remarks

Controllers and the piloting community have been involved in the design throughout the program. In the case of FAST, most operational issues could be adequately addressed through a combination of real-time simulation and shadow-mode testing. Operational tests are scheduled to begin this fall to validate the concept in real operations in anticipation of national deployment. In the case of DA, the total system performance is highly dependent on the compatibility between aircraft or pilot and controller procedures. Issues that will affect system performance include the adequacy of the aircraft and wind modeling, and the ability and willingness of the crew to follow DA advisories. This difference has led to a greater involvement by pilots throughout the design and the initiation of early and non-intrusive field evaluations.

Fast-time simulations and analysis of real-time data are used to quantify the performance of the system and to provide a basis for extrapolating limited results from real-time simulation, shadow-mode testing, and limited field tests to a variety of cases in a statistically significant manner.

Results to date indicate a tremendous operational benefit through the introduction of CTAS type automation tools.

References

[1] Erzberger, H.; Davis, T. J.; and Green, S. M.: *Design of Center-TRACON Automation System.* Proceedings of the AGARD Guidance and Control Panel 56th Symposium on Machine Intelligence in Air Traffic management, Berlin, Germany, 1993, pp. 52-1–52-14.

[2] Nedell, W.; and Erzberger, H.: *The Traffic Management Advisor.* Proceedings of the American Control Conference, San Diego, Calif., May 1990.

[3] Green, S. M.: *Time-Based Operations in an Advanced ATC Environment.* Proceedings of the Aviation Safety/Automation Program Conference, NASA CP-3090, Virginia Beach, Va., Oct. 1989, pp. 249–260.

[4] Davis, T. J.; Krzeczowski, K. J.; and Bergh, C. C.: *The Final Approach Spacing Tool.* Procccdings of thc 13th IFAC Symposium on Automatic Control in Aerospace, Palo Alto, Calif., Sept. 1994.

[5] Lee, K. K.; and Davis, T. J.: *The Development of the Final Approach Spacing Tool (FAST): A Cooperative Controller-Engineer Design Approach.* Proceedings of the 14th IFAC Symposium on Automatic Control in Aerospace, Berlin, Germany, Sept. 1995.

[6] Krzeczowski, K. J.; Davis, T. J.; Erzberger, H.; Lev-Ram, I.; and Bergh, C. P.: *Knowledge-Based Scheduling of Arrival Aircraft in the Terminal Area.* Proceedings of the AIAA Guidance, Navigation, and Control Conference, Baltimore, Md., Aug. 1995.

[7] Slattery, R. A.: *Terminal Area Trajectory Synthesis for Air Traffic Control Automation.* Conference Proceedings of the American Control Conference, June 1995.

[8] Davis, T. J.; Erzberger, H.; Green, S. M.; and Nedell, W.: Design and Evaluation of an Air Traffic Control Final Approach Spacing Tool. *Journal of Guidance, Control, and Dynamics*, vol. 14, no. 4, July–Aug. 1991, pp. 848–854.

[9] Brinton, C. R.: *An Implicit Enumeration Algorithm for Arrival Aircraft Scheduling.* Proceedings of the 11th Digital Avionics Systems Conference, Seattle, Wash., Oct. 1992.

[10] Cooper, G. E.; and Harper, R. P.: *The Use of Pilot Rating in the Evaluation of Aircraft Handling Qualities*. NASA TN D-5153, 1969.

[11] Ballin, M. G.; and Erzberger, H.: *An Analysis of Aircraft Landing Rates and Separations at Dallas/Ft. Worth Airport*. NASA TM-110397, July 1996.

[12] Neuman, F.; and Erzberger, H.: *Analysis of Sequencing and Scheduling Methods for Arrival Traffic*. NASA TM-102795, April 1990.

[13] Erzberger, H.: *Integrating Physical Models and Expert Knowledge in the Design of Automated Air Traffic Management Systems*. AGARD Lecture Series No. 200, Knowledge Based Functions in Aerospace Systems, Nov. 1995.

[14] Neuman, F.; Erzberger, H.; and Schuellar, M. S.: *CTAS Data Analysis Program*. NASA TM-108842, Ames Research Center, 1994.

[15] Green S. M.; Vivona, R. A.; and Sanford, B.: *Descent Advisor Preliminary Field Test*. Proceedings of the AIAA Guidance, Navigation, and Control Conference, Baltimore, Md., Aug. 7–9, 1995.

SCHEDULING MODELS AND ALGORITHMS FOR TMA TRAFFIC MANAGEMENT

Lucio Bianco

Dipartimento di Informatica, Sistemi e Produzione
University of Rome "Tor Vergata"
Viale della Ricerca Scientifica – 00133 Rome, Italy
I.A.S.I. - C.N.R. – Viale Manzoni, 30 – 00185 Rome, Italy

Paolo Dell'Olmo

Dipartimento di Informatica, Sistemi e Produzione
University of Rome "Tor Vergata"
Viale della Ricerca Scientifica – 00133 Rome, Italy
I.A.S.I. - C.N.R. – Viale Manzoni, 30 – 00185 Rome, Italy

Stefano Giordani

Dipartimento di Informatica, Sistemi e Produzione
University of Rome "Tor Vergata"
Viale della Ricerca Scientifica – 00133 Rome, Italy

In this paper, models and algorithms for real-time control of the TMA are proposed. We consider two cases: in the first one (static) we assume that there is a set of aircraft to be sequenced for which we know in advance their entry time in the terminal area; in the second one (dynamic), the entry times of future aircraft arc unknown and the sequence of aircraft is recomputed whenever a new aircraft approaches the terminal area. For the static case, we model the sequencing problem as a Cumulative Traveling Salesman Problem with Ready Times and propose two lower bounds for testing heuristic solutions. For the dynamic case, where only a limited knowledge of the arrivals is assumed, we add to the basic model a set of constraints which allow the controller to maintain given patterns of the landing sequences previously generated. For both cases, heuristic algorithms are proposed and computational results are discussed.

1. Introduction and Background

One of the main problems Air Traffic Control (ATC) has to face nowadays is flight delay caused by air traffic congestion. To a certain extent this is due to the traffic distribution that, owing to the requirements of transport users, is mainly concentrated at large airports. Then, it becomes critical the management of the region around the airport (i.e., the terminal area (TMA)). This requires consideration of various functions: approach geometry, on-line control of nominal trajectories, path-stretching and holding maneuvers, metering, sequencing and spacing the aircraft, conflict alert and resolution and any other emergency function, runway capability and their ways of operation according to weather conditions.

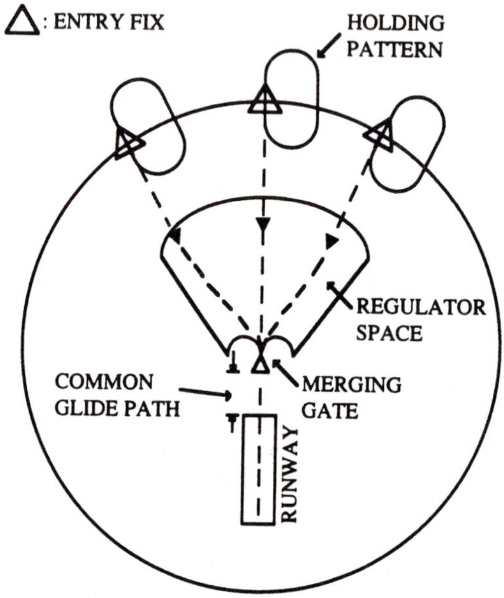

Figure 1. Structure of TMA

During peak traffic periods, the control of an aircraft arrivals in TMA becomes a very complex task. The operations on final approach of air traffic controllers, among other aspects, must guarantee that every aircraft waiting to land maintains the required degree of safety. They also have to decide what aircraft should use a particular runway, at what time this should be done and what maneuvering should be executed to achieve this. The viable accomplishment of such a task becomes more difficult in view of the fact that, during peak periods, the demand for runway occupancy may reach or even exceed the capabilities of the system. Under such bottleneck conditions, an increase in collision risk can logically be expected and, because of safety considerations, the structure of TMA is rigidly defined and all aircraft must fly in a manner satisfying prefixed procedural constraints.

To simplify the understanding of the problem we refer to an idealized representation of TMA as shown in Figure 1, considering only landings.

The following aspects must be underlined:

- every aircraft must approach the runway for landing, flying along one of the prestructured paths of TMA;
- the runway can be occupied by only one aircraft at a time;
- every aircraft must fly along the common approach path following a standard descent profile;
- during all the approach phases a separation standard between every pair of consecutive aircraft must be maintained;
- the sequencing strategy used by almost all major airports of the world is the First-Come First-Served (FCFS) discipline.

As it is well known, FCFS strategy is simple to implement but is likely to produce excessive delays. Therefore, an effort must be made to minimize the delay or optimize some other measure of performance related to passengers discomfort and other costs, without violating safety constraints.

Consequently, the TMA problem can be stated as follows: Given a set of aircraft entering the TMA and given, for each aircraft the Preferred Landing Time (PLT), the runway occupancy time, the cost per unit time of flight, the geometry of the approach path and glide path and the corresponding aircraft speeds, assign to each aircraft the starting time from the fix and the approach path in such a way that the procedural constraints are satisfied and a given system performance index is optimized. With the TMA operating in the aforementioned way, the TMA problem can be decomposed into the two following subproblems:

1) given the constraints on aircraft performance, the initial and final states (position and speed) and the pre-established flight time, determine the optimum trajectories which connect these states with the specified flight time;

2) given a set of Preferred Landing Times (PLT), determine the Actual Landing Times (ATL) sequence which satisfies the procedural constraints on the runway and the glide path and optimizes a system performance index.

To a large extent, these two problems are independent. In fact, the required controls to follow the approach paths can be calculated in advance. Therefore, the need of real time calculations is limited only to subproblem 2, denoted as Aircraft Sequencing Problem (ASP).

There are two important aspects of ASP we want to underline. First, safety regulations state that any two coaltitudinal aircraft must maintain a "minimum horizontal separation", which is a function of the type and of the relative positions of the two aircraft . Second, the "landing speed" of any type of aircraft is generally different from the landing speed of another aircraft type. As a consequence of the variability of the above parameters the *Landing Time Interval* (LTI), which is the minimum permissible time interval between two successive landings, is a variable quantity. These differences in separation are mandatory and recognized by federal regulations. Then, it is easy to realize that a sequence may take a longer time to land a set of aircraft if there must be large time gaps between them because of safety reasons, while another sequence may require a shorter time if it eliminates the long gaps by taking advantage of the asymmetries in the LTIs. For this reason the sequence in which the aircraft are ready to land (i.e., the FCFS sequence) might not be a good one with respect to optimization. Moreover, as the TMA is a dynamic system, a better sequencing requires a continuous rearranging of a good sequence. This can be accomplished only taking into account practical constraints, like, for instance ,the maximum delay time for a given aircraft. Hence, the sequences cannot be arbitrarily reordered.

The Aircraft Sequencing Problem has been studied in a number of papers (see [1,2,7,11,12,16]). In [7], Dear carries out an excellent investigation of the ASP. In particular, the author points out that, in order to determine the landing sequence, we need to consider all aircraft in the system. This is denoted as the "static" version of the problem where all aircraft are present at the same time in the holding stacks and they can land at any time. In the "dynamic" version, the author considers the composition of the aircraft mix to change over time. The operational constraints on the rearrangements of the sequence mentioned above are modeled introducing the notion of Constrained Position Shifting (CSP). With respect to this, the final position of an aircraft in the actuated sequence cannot differ from the initial one by a specified parameter, termed Maximum Position Shifting (MPS). In [11,12], Psaraftis develops an exact dynamic programming algorithm for the static case which implements also the CPS concept, with the objective of minimizing the total time required to land a set of aircraft having the same PLT. The sequencing problem is recognized to have a structure similar to the Traveling Salesman Problem (TSP). Grouping aircraft by weight-class, and considering that the number of different classes is fixed, it is possible to show that the algorithm runs in polynomial time. In [5], referring to the same optimization criterion, a combinatorial approach is proposed which, for selecting the aircraft to land next, takes into account not only the weight class but also the Preferred Landing Time of the individual aircraft with the same optimization criterion. In [16], Venkatakrishan, Barnett and Odoni develop three models, and corresponding algorithms, for the Boston Logan Airport terminal airspace and apply, on real airflow data, sequencing algorithms to expedite the landing of incoming aircraft. The paper reports results on accurate data sets for the Logan airport which show that better sequencing can reduce delays by 30% in some instances.

In this paper, we first study a combinatorial optimization model for the ASP for the case in which the objective is the minimization of the average aircraft arrival delay of a set of aircraft, for which the entry time and the weight class are known. For this case, we give a new mathematical formulation of the problem, which allows to define some lower bounds of the optimal solution and the design of a fast heuristic algorithm. Performances of the heuristic are measured, with respect to the lower bounds of the problem, in scenarios with different traffic volumes. Then we present a model for the case in which the knowledge of incoming aircraft is limited. The proposed model is extended in order to include some operational constraints which are represented by Maximum Position Shifting rule for individual aircraft and Relative Position Shifting rule for subsequences of the current landing sequence. The proposed heuristic is fast enough to be reasonably applied in a closed control loop approach, which recomputes in real time the actual sequence whenever the system is perturbed (arrival of a new aircraft, change of constraints, delays, etc.). Computational results for different traffic volumes are discussed.

2. A Machine Scheduling Model for the ASP

Suppose that the air traffic controller is confronted with the following problem. A number n of aircraft are waiting to land at different PLT at a single runway airport. His task is to find a landing sequence for this set of aircraft so that a certain measure of performance is optimized, while all problem constraints are satisfied.

We make the following assumptions:

a) The pilots of all aircraft are capable and willing to execute the instructions of the controller, given enough prior notice.

b) The measure of performance is the cumulative cost of delays of aircraft. In particular, we consider the sum of the time differences between an aircraft's completion landing time (i.e., the sum of the actual landing time and the time required to leave the runway) and its preferred landing time.

c) The satisfaction of the Landing Time Interval constraints is required. This means that the time interval between the landing of an aircraft j, followed by the landing of an aircraft i, must not be less than a known time interval c_{ji}.

d) In the first model (static), we assume the composition of the set of aircraft completely known (this assumption will be removed in the dynamic model in Section 7) and that the set of LTIs is known.

e) At any stage of the sequencing procedure, the air traffic controller is free to assign the next landing slot to any of the remaining aircraft. This means that the initial position which the aircraft had when arrived at TMA is ignored. This assumption will be removed in the dynamic model in Section 7.

At this point, it is not difficult to see that the above problem can be represented as a particular machine scheduling problem. In fact, with the aforementioned assumptions, the following analogy can be established:

- to each landing operation is associated a job;
- the runway corresponds to a machine with capacity one;
- the PLT_j of aircraft j corresponds to the ready time r_j of job j;
- the ALT_j of aircraft j corresponds to the start time s_j of job j;

- the time the aircraft j leaves the runway corresponds to the completion time C_j of job j;

- the minimum time interval between the landing of aircraft j followed by the landing of aircraft i, corresponds to the sequence dependent processing time c_{ji}.

Therefore the ASP problem, as defined here, can be reformulated, using the machine scheduling notation proposed by Graham *et al.* [10], as $1|r_j$, seq-dep$|\Sigma C_j$.

The problem of scheduling with ready times and sequence dependent processing times on a single machine to minimize the sum of completion times can be stated as follows. A set $J = \{j|j = 1, 2,..., n\}$ of n jobs has to be processed on a single machine. For each job j, the ready time r_j is given, while the processing time is dependent on the job position in the sequence. Then, we denote with c_{ji}, the time required by the machine for processing job j immediately before job i. Let be denoted by C_j the completion time of the job j, the problem consists in finding a schedule such that the total completion time $\sum_{j=1}^{n} C_j$ is minimum.

The problem of sequencing n jobs on one machine has been studied extensively under different assumptions and objective functions. In the case in which sequence dependent processing times are considered, at the best of our knowledge, only the problem with equal ready times and equal weights (i.e. $1|$seq-dep$|\Sigma C_j$) has been studied. The problem, proved to be NP-hard by Rinnooy Kan [14], was also studied by Bianco, Mingozzi and Ricciardelli [4], who have shown that it is equivalent to the Cumulative Traveling Salesman Problem (CTSP) and proposed exact and heuristic solution algorithms; Fischetti, Laporte and Martello [8] have also given another formulation and another exact algorithm.

For application to the ASP, we study the extension of the $1|$seq-dep$|\Sigma C_j$ scheduling problem to the case in which unequal ready times are given. We state that the ASP, as defined above, is equivalent to a special case of the Cumulative Traveling Salesman Problem with additional time constraints in which each city v_j is ready to be visited at time r_j. For related results on machine scheduling problems and technical details regarding the model $1|r_j$, seq-dep$|\Sigma\, C_j$ the reader can see [3].

3. Notation and Definitions

Without loss of generality, let us assume that the job $j_1 \in J$ is a dummy job and consider only the feasible schedules starting with it; in fact, the dummy job represents the "initial" state of the machine before the job processing begins. Let us indicate with $S = (j_{i_1},\ldots, j_{i_k}, j_{i_{k+1}},\ldots, j_{i_n})$ a schedule (i.e. a permutation) of the job set J, with $j_{i_1} \equiv j_1$. The completion time C_{i_k} of the job j_{i_k} occupying position k in the schedule S is given by $C_{i_k} = (s_{i_k} + c_{i_k i_{k+1}})$, where s_{i_k} is the starting time of the job j_{i_k}, and assuming $c_{i_n i_{n+1}} \equiv c_{i_n i_1}$. Obviously, s_{i_k} depends on the ready time r_{i_k} of the job j_{i_k} and on the completion time $C_{i_{k-1}}$ of the previous scheduled job $j_{i_{k-1}}$, that is $s_{i_k} = \max\,[r_{i_k}, C_{i_{k-1}}]$. Defining with $\delta_{i_k} = \max\,[0, r_{i_k} - C_{i_{k-1}}]$ the machine idle time before starting to process job j_{i_k}, it results $C_{i_k} = (C_{i_{k-1}} + \delta_{i_k} + c_{i_k i_{k+1}})$. Assuming that $j_{i_{n+1}} \equiv j_1$, that is at the end of the job processing the machine has to be in the "initial" state, the cost of the scheduling S, given by $z(S) = \sum_{k=1}^{n} C_{i_k}$, may be written as

$$z(S) = \sum_{k=1}^{n} \sum_{h=1}^{k} (\delta_{i_h} + c_{i_h i_{h+1}})$$

that is

$$z(S) = \sum_{k=1}^{n} (n-k+1)(\delta_{i_k} + c_{i_k i_{k+1}})\,.$$

The problem is finding a schedule S starting with job j_1 that minimize $z(S)$.

We show that the $1|r_j$, seq-dep$|\Sigma C_j$ scheduling problem consists in finding a hamiltonian tour H in a complete directed graph G. Let us consider the complete directed graph $G = (V, A)$, where the set of vertices $V = \{v_1,..., v_n\}$ represents the set of jobs J and each arc $(j, i) \in A$ identifies the choice of scheduling job j immediately before job i; a cost c_{ji}, equal to the sequence dependent processing time of the job j respect to the job i, is associated to the arc (j, i) and represents the time that has to be spent to visit v_j and to travel from v_j to v_i; for each vertex v_j a ready time r_j, from that forth the vertex is ready to be visited, is also given. Therefore, a feasible schedule S is related to an hamiltonian tour $H = (v_{i_1},..., v_{i_n}, v_{i_{n+1}})$ in G, where $v_{i_1} \equiv v_{i_{n+1}} \equiv v_1$ (vertex v_{i_1} occupies position 1 and $n+1$), subject to visit each vertex v_{i_k} at a time s_{i_k} not less than its ready time r_{i_k}.

Let us consider a cumulative cost $z(H)$ of the traveled time distance for the tour H, defined as follows. Let l_k be the cost of the time distance traveled to go from vertex v_1 to vertex v_{i_k}, occupying position k in the tour H; it results

$$
\begin{aligned}
&l_1 = 0, \quad l_2 = \delta_{i_1} + c_{i_1 i_2}, \quad l_3 = (\delta_{i_1} + c_{i_1 i_2}) + (\delta_{i_2} + c_{i_2 i_3}), \quad ..., \\
&l_k = (\delta_{i_1} + c_{i_1 i_2}) + (\delta_{i_2} + c_{i_2 i_3}) + ... + (\delta_{i_{k-1}} + c_{i_{k-1} i_k}), \quad ..., \\
&l_{n+1} = (\delta_{i_1} + c_{i_1 i_2}) + (\delta_{i_2} + c_{i_2 i_3}) + ... + (\delta_{i_{k-1}} + c_{i_{k-1} i_k}) + ... + \\
&\qquad (\delta_{i_n} + c_{i_n i_{n+1}}),
\end{aligned}
$$

where δ_{i_k} is the time that has to be waited at vertex v_{i_k} before it is ready to be visited. Then, the cumulative cost $z(H)$ of tour H is given by

$$z(H) = \sum_{k=1}^{n} l_{k+1} = \sum_{k=1}^{n} \sum_{h=1}^{k} (\delta_{i_h} + c_{i_h i_{h+1}}),$$

where it may be noted that the arc (v_{i_1}, v_{i_2}) in position 1 gives the contribution $n(\delta_{i_1} + c_{i_1 i_2})$, arc (v_{i_2}, v_{i_3}) in position 2 gives the contribution $(n - 1)(\delta_{i_2} + c_{i_2 i_3})$, arc $(v_{i_k}, v_{i_{k+1}})$ in position k gives the contribution $(n - k + 1)(\delta_{i_k} + c_{i_k i_{k+1}})$ and, finally, arc $(v_{i_n}, v_{i_{n+1}})$ in position n gives the contribution $(\delta_{i_n} + c_{i_n i_{n+1}})$.

Hence, the cost function $z(H)$ may also be written as

$$z(H) = \sum_{k=1}^{n}(n-k+1)(\delta_{i_k} + c_{i_k i_{k+1}}),$$

which has the same expression as $z(S)$, the cost of the schedule S related to the tour H.

Therefore, the $1|r_j$, seq-dep$|\Sigma C_j$ scheduling problem is equivalent to the problem of finding a minimum cumulative cost hamiltonian tour in a graph G, satisfying a set of (visiting) time constraints, and in which the cost also takes in account the waiting times spent at vertices which are not ready to be visited. We refer to this problem as the Cumulative Traveling Salesman Problem with Ready Times (CTSP-RT).

4. Problem Formulation

In this section, we give a formulation of the CTSP-RT. Let us consider a position dependent TSP variable [13], that is a binary decision variable x_{ij}^k such as $x_{ij}^k = 1$ if the arc (v_i, v_j) is in position k in a hamiltonian tour H, and $x_{ij}^k = 0$ otherwise. Furthermore, let us consider a non-negative real variable δ_i^k, representing the waiting time spent by the salesman at vertex v_i, before visiting it as the kth vertex in the tour H. If arc (v_i, v_j) is in position k, its contribution to the cost function is given by $(n - k + 1)(\delta_i^k + c_{ij})$.

The CTSP-RT may be formulated in terms of mixed integer programming formulation as follows:

(CTSP-RT)

$$\min z = \sum_{k=1}^{n} (n-k+1) \sum_{i=1}^{n} (\delta_i^k + \sum_{j=1}^{n} c_{ij} x_{ij}^k) \qquad (1)$$

s.t.

$$\sum_{j=1}^{n} x_{ij}^k - \sum_{l=1}^{n} x_{li}^{k-1} = 0, \qquad i = 2, \ldots, n, \; k = 2, \ldots, n \qquad (2)$$

$$\sum_{j=1}^{n} x_{1j}^1 = 1 \qquad (3)$$

$$\sum_{i=1}^{n} x_{i1}^{n} = 1 \qquad (4)$$

$$\sum_{k=1}^{n}\sum_{i=1}^{n} x_{ij}^{k} = 1, \qquad j = 1,\dots,n \qquad (5)$$

$$\sum_{k=1}^{n}\sum_{j=1}^{n} x_{ij}^{k} = 1, \qquad i = 1,\dots,n \qquad (6)$$

$$\sum_{i=1}^{n}\sum_{j=1}^{n} x_{ij}^{k} = 1, \qquad k = 1,\dots,n \qquad (7)$$

$$r_i \sum_{j=1}^{n} x_{ij}^{1} = \delta_i^{1}, \qquad i = 1,\dots,n \qquad (8)$$

$$r_i \sum_{j=1}^{n} x_{ij}^{k} \leq \delta_i^{k} + \sum_{h=1}^{k-1}\sum_{l=1}^{n}(\delta_l^{h} + \sum_{j=1}^{n} c_{lj} x_{lj}^{h}), \qquad i = 1,\dots,n,\ k = 2,\dots,n \qquad (9)$$

$$x_{ij}^{k} \in \{0,1\}, \qquad \forall k, \forall (i,j) \qquad (10)$$

$$\delta_i^{k} \geq 0, \qquad \forall k, \forall i. \qquad (11)$$

Constraints (2) ensure continuity in each vertex. Constraints (3) and (4) impose that each tour must start from vertex v_1, with an arc (v_1, v_j) in position 1, and must end in the same vertex, with an arc (v_i, v_1) in position n. Constraints (5) and (6) state that each vertex in the tour must be visited once and only once. Constraints (7) impose that in each position k there is exactly one arc. It may be noted that the set of constraints (7) together with one of the two sets (5) and (6) are redundant, but this redundancy will be exploited to get the relaxation proposed in Section 5. Finally, constraints (8) and (9) impose a non-zero waiting time δ_i^k that has to be waited at the kth vertex v_i in the tour before it is ready to be visited.

5. Lower Bounds

Now, we describe two lower bounds for the optimal solution value of the $1|r_j$, seq-dep$|\Sigma C_j$ scheduling problem. In the first lower bound, denoted by LB1, we consider a relaxation of the ready times, by fixing every non dummy job $j \in J\setminus\{1\}$ to be ready to be visited at time $r = \min_{j \in J\setminus\{1\}}[r_j]$; in the second one, named LB2, we relax the sequence dependent processing time, by considering for each job a sequence independent change over times.

LB1 is a lower bound based on a Lagrangian relaxation of the CTSP-RT mathematical model (1)–(11).

Let us dualize constraints (5) and (6) and drop constraints (9). The relaxed problem is

(RP)

$$LR(\lambda) = \min \sum_{i,k=1}^{n} (n-k+1)\delta_i^k + \sum_{i,j,k=1}^{n} [(n-k+1)c_{ij} + \lambda_i + \lambda_j] x_{ij}^k - 2\sum_{i=1}^{n} \lambda_i$$

subject to constraints (2), (3), (4), (7), (8), (10), (11).

Every feasible solution to RP is a tour T of cardinality n, as stated by constraints (2) and (7), that starts and ends at vertex v_1, as imposed by constraints (3) and (4); moreover, the tour starts at time r_1 as for constraints (8). Note that a feasible tour is not necessarily elementary, in the sense that it may include loops formed by some consecutive vertices. In fact, whenever constraints (5) and/or (6) are violated, in the subgraph constituted by the tour, some vertices may have degree d_i greater than 2 and some others have d_j equal to 0.

The lower bound LB1 of CTSP-RT may be computed as

$$\text{LB1} = \max_{\lambda}[LR(\lambda)] + (n-1)\bar{\delta}_2,$$

where $\bar{\delta}_2$ is a lower bound for the time that has to be waited before visiting the vertex in position 2 in the tour T, that is $\bar{\delta}_2 = \max\{0, \min_{i \in V \setminus \{v_1\}}[r_i - (r_1 + c_{1i})]\}$.

The computation of $\max_{\lambda}[LR(\lambda)]$ may be accomplished by means of subgradient technique [9], that is updating the penalties $\{\lambda_i\}$ as

$$\lambda_i = \lambda_i + \theta \frac{(z_{\mathrm{H}} - LR(\lambda))(d_i - 2)}{\sum_{j=1}^{n} (d_j - 2)^2}$$

at each iteration, where z_H is an upper bound (i.e. a heuristic solution value) on the optimal solution of CTSP-RT, and θ a scalar satisfying $0 < \theta \le 2$, starting with $\theta = 2$ and halving its value whenever $LR(\lambda)$ has failed to increase in some fixed number of iterations; a natural choice for the initial Lagrangian multipliers is zero.

Now, let us show how to solve RP. As explained before, RP corresponds to the problem of finding a minimum cost tour of cardinality n, which may be conveniently solved by dynamic programming as shown in the sequel.

Let $h(k, v_i)$ be the cost of the minimum cost path of cardinality k, that starts at vertex v_i and ends in v_1. The function $h(k, v_i)$ may be recursively computed from

$$h(k, v_i) = \min_{v_j \in V \setminus \{v_i\}} [h(k-1, v_j) + c_{ij}^k],$$

with

$$h(0, v_i) = \begin{cases} 0 & \text{if } v_i = v_1 \\ \infty & \text{otherwise,} \end{cases}$$

where $c_{ij}^k = k\, c_{ij} + \lambda_i + \lambda_j$.

Therefore,

$$LR(\lambda) = n\, r_1 + h(n, v_1) - 2\sum_{i=1}^{n} \lambda_i.$$

The tour corresponding to $LR(\lambda)$ is not necessarily an elementary tour and it may also contains sub-cycles of cardinality 2 (i.e. $v_r v_s v_r$). The bound may be improved imposing that any sub-cycle should contain at least three different vertices [6].

Let us propose another lower bound based on a relaxation of the scheduling model. The basic idea is to relax the original problem into another one, considering for each job j a processing time $q_j = \min_{i \in J \setminus \{j\}}[c_{ij}]$ that does not depend on the sequence. This latter problem is equivalent to the $1|r_j|\Sigma C_j$ problem with processing times q_j, and provides a lower bound on the former since $q_j \le c_{jj}$ for any $i \in J \setminus \{j\}$. Since $1|r_j|\Sigma C_j$ is a strongly NP-hard problem, we make another relaxation allowing job preemption. As well known, this problem can be optimally solved in $O(n \log n)$ by the Shortest Remaining Processing Time rule [15]. Finally, the returned value gives the lower bound LB2 on the $1|r_j, \text{seq-dep}|\Sigma C_j$ problem.

6. Heuristic Algorithm and Computational Results

In this section, we present a heuristic algorithm for the $1|r_j, \text{seq-dep}|\Sigma C_j$ scheduling problem, denoted by CIH, which is a greedy heuristic inspired to the *cheapest insertion* heuristic for the TSP. CIH finds a feasible schedule, starting with the dummy job $1 \in J$, in n-1 iterations.

At each iteration, the best partial schedule obtained inserting an unscheduled job in the sequence of the scheduled ones is selected.

Let us consider iteration k+1 and let S_k be the partial schedule corresponding to the sequence $(j_1,..., j_k)$ of the k jobs selected in the previous k iterations. We seek for a schedule S_{k+1} of k+1 jobs of minimum total completion time, chosen between the set of all the feasible schedules obtained from S_k inserting an unscheduled job $j \in \bar{S}_k$ in all the possible positions in the sequence $(j_1,..., j_k)$ related to S_k.

A more detailed description of the heuristic is as follows:

CIH Algorithm

Step 1. Let $S_1 := \{1\}$, $U := J \backslash \{1\}$, $k := 1$.

Step 2. While $U \neq \Phi$, do

- **Step 2.1.** Select a job $j \in \bar{S}_k$ such that it can be inserted in the sequence related to S_k at the lowest increase cost; let h be the relative insertion position;
- **Step 2.2.** Insert j at position h in the sequence related to S_k and let S_{k+1} be the new feasible schedule related to the new sequence;
- **Step 2.3.** Let $U := U \backslash \{j\}$, $k := k + 1$.

The heuristic algorithm and the lower bound procedures were implemented in C programming language on a Sun Sparcstation 1+.

To test the efficiency of the algorithms when applied to real-world problems we must consider that aircraft, waiting to land, can be classified into a relatively small number of distinct categories, according to speed, capacity, weight and other technical characteristics. Hence, the minimum LTI between two successive aircraft is a function only of the categories they belong to. Table 1 shows the minimum LTIs relative to main categories of commercial aircraft. In particular, we consider $m = 4$ categories: category number 1 identifies Boeing 747 (B747), category number 2 corresponds to Boeing 727 (B727), category number 3 identifies Boeing 707 (B707) and finally category number 4 corresponds to Mc Donnel Douglas DC9 (DC9).

i \ j	1	2	3	4
1	96	200	181	228
2	72	80	70	110
3	72	100	70	130
4	72	80	70	90

1 = B747; 2 = B727; 3 = B707; 4 = DC9

Table 1. Minimum landing time intervals: c_{ij} (sec.); m=4

Tables 2, 3 illustrate the results of two realistic large scale problems with 30 and 44 aircraft, respectively, while Tables 4, 5 show the results for the case of 30 aircraft in which we simulated a less and a more congested case, respectively. In these tables, in the first three columns we list the aircraft identification numbers, categories and preferred landing times; in the fourth column we give the actual landing time issued by the FCFS discipline; in the last three columns we list the sequence of landing aircraft, with their categories and actual landing times, issued by the CIH algorithm. Furthermore, the mean and maximum delays (i.e., the difference between actual and preferred landing times) and the total (landing) completion time of the aircraft for the FCFS discipline (z_{FCFS}) and for the CIH algorithm

(z_{CIH}) are given; performance values are also reported. It can be noted that in all the considered case the CIH algorithm performs better than FCFS discipline.

			FCFS	CIH		
Airc. Num.	Cat.	PLT (sec)	ALT (sec)	Airc. Seq.	Cat.	ALT (sec)
1	1	0	0	1	1	0
2	1	79	96	2	1	96
3	1	144	192	3	1	192
4	2	204	392	5	1	288
5	1	264	464	6	1	384
6	1	320	560	7	2	584
7	2	528	760	4	2	664
8	1	635	832	9	2	744
9	2	730	1032	10	2	824
10	2	766	1112	11	1	896
11	1	790	1184	12	1	992
12	1	920	1280	8	1	1088
13	3	1046	1461	17	2	1288
14	4	1106	1591	16	2	1368
15	2	1136	1671	15	2	1448
16	2	1166	1751	14	4	1558
17	2	1233	1831	13	3	1628
18	1	1642	1903	20	3	1770
19	1	1715	1999	18	1	1842
20	3	1770	2180	19	1	1938
21	1	2074	2252	21	1	2074
22	1	2168	2348	22	1	2170
23	4	2259	2576	23	4	2398
24	2	2427	2656	24	2	2478
25	1	2481	2728	25	1	2550
26	2	2679	2928	26	2	2750
27	3	2883	2998	27	3	2883
28	2	2982	3098	28	2	2983
29	1	3046	3170	29	1	3055
30	1	3091	3266	30	1	3151

FCFS *discipline*
Mean Delay = 267.6 sec
Max Delay = 598.0 sec
z_{FCFS} = 53577.0 sec

CIH *algorithm*
Mean Delay = 126.6 sec
Max Delay = 582.0 sec
z_{CIH} = 49060.0 sec

Lower bounds
LB1 = 38400.8 sec
LB2 = 42284.0 sec

Performance values
$100(z_{CIH}\text{-LB2})/\text{LB2}=16.0\%$
$100(z_{FCFS}-z_{CIH})/z_{CIH}=9.2\%$

Table 2. Computational results on a realistic ASP instance (n = 30)

Airc. Num.	Cat.	PLT (sec)	FCFS ALT (sec)	CIH Airc. Seq.	CIH Cat.	CIH ALT (sec)
1	1	0	0	1	1	0
2	1	79	96	2	1	96
3	2	144	296	5	2	296
4	2	204	376	4	2	376
5	2	264	456	3	2	456
6	1	320	528	7	1	528
7	1	528	624	6	1	624
8	1	635	720	8	1	720
9	2	730	920	10	1	816
10	1	766	992	12	1	920
11	2	790	1192	14	2	1120
12	1	920	1264	16	2	1200
13	2	1046	1464	17	2	1280
14	2	1106	1544	19	2	1360
15	1	1136	1616	20	2	1440
16	2	1166	1816	13	2	1520
17	2	1226	1896	11	2	1600
18	1	1233	1968	9	2	1680
19	2	1286	2168	23	2	1760
20	2	1418	2248	25	2	1840
21	1	1642	2320	26	2	1920
22	1	1715	2416	28	2	2000
23	2	1749	2616	27	2	2080
24	1	1770	2688	29	1	2152
25	2	1809	2888	30	1	2248
26	2	1869	2968	32	1	2344
27	2	1929	3048	34	1	2440
28	2	1989	3128	35	1	2536
29	1	2074	3200	24	1	2632
30	1	2168	3296	39	1	2728
31	2	2229	3496	22	1	2824
32	1	2259	3568	40	1	2920
33	2	2326	3768	41	1	3016
34	1	2427	3840	43	1	3112
35	1	2481	3936	42	1	3208
36	2	2488	4136	21	1	3304
37	2	2565	4216	15	1	3400
38	2	2657	4296	18	1	3496
39	1	2679	4368	44	2	3696
40	1	2883	4464	38	2	3776
41	1	2982	4560	37	2	3856
42	1	3046	4656	36	2	3936
43	1	3091	4752	33	2	4016
44	2	3153	4952	31	2	4096

FCFS *discipline*

Mean Delay = 881.0 sec
Max Delay = 1799.0 sec
z_{FCFS} = 114712.0 sec

CIH *algorithm*

Mean Delay = 463.4 sec
Max Delay = 2264.0 sec
z_{CIH} = 95456.0 sec

Lower bounds

LB1 = 82018.9 sec
LB2 = 70977.0 sec

Performance values

$100(z_{CIH}\text{-}LB2)/LB2=16.4\%$
$100(z_{FCFS}\text{-}z_{CIH})/z_{CIH}=20.2\%$

Table 3. Computational results on a realistic ASP instance (n = 44)

Airc. Num.	Cat.	PLT (sec)	FCFS ALT (sec)	CIH Airc. Seq.	CIH Cat.	CIH ALT (sec)
1	1	0	0	1	1	0
2	1	95	96	2	1	96
3	1	137	192	3	1	192
4	2	305	392	4	2	392
5	1	621	621	5	1	621
6	1	684	717	6	1	717
7	2	797	917	7	2	917
8	1	941	989	9	2	1088
9	2	1088	1189	8	1	1160
10	2	1308	1308	10	2	1360
11	1	1347	1380	11	1	1432
12	1	1511	1511	12	1	1528
13	3	2026	2026	13	3	2026
14	4	2155	2156	14	4	2156
15	2	2512	2512	15	2	2512
16	2	2719	2719	16	2	2719
17	2	2804	2804	17	2	2804
18	1	3304	3304	18	1	3304
19	1	3112	3130	19	1	3130
20	3	3249	3311	20	3	3311
21	1	3304	3383	21	1	3383
22	1	3375	3479	22	1	3479
23	4	3474	3707	23	4	3679
24	2	3650	3787	24	2	3789
25	1	3740	3859	26	2	3898
26	2	3898	4059	27	3	3968
27	3	3916	4129	28	2	4084
28	2	4084	4229	25	1	4156
29	1	4119	4301	29	1	4252
30	1	4230	4397	30	1	4348

FCFS *discipline*

Mean Delay = 70.0 sec
Max Delay = 233.0 sec
z_{FCFS} = 73600.0 sec

CIH *algorithm*

Mean Delay = 66.5 sec
Max Delay = 219.0 sec
z_{CIH} = 73403.0 sec

Lower bounds

LB1 = 38403.5 sec
LB2 = 68235.0 sec

Performance values

$100(z_{CIH}\text{-LB2})/\text{LB2} = 7.6\%$
$100(z_{FCFS}-z_{CIH})/z_{CIH} = 0.3\%$

Table 4. Computational results on a realistic ASP instance (n = 30, uncongested case)

Airc. Num.	Cat.	PLT (sec)	FCFS ALT (sec)	CIH Airc. Seq.	CIH Cat.	CIH ALT (sec)
1	1	0	0	1	1	0
2	1	54	96	2	1	96
3	1	118	192	3	1	192
4	2	140	392	5	1	288
5	1	282	464	8	1	384
6	1	297	560	6	1	480
7	2	308	760	12	1	576
8	1	366	832	11	1	672
9	2	383	1032	17	2	872
10	2	409	1112	16	2	952
11	1	507	1184	15	2	1032
12	1	531	1280	10	2	1112
13	3	567	1461	9	2	1192
14	4	673	1591	26	2	1272
15	2	690	1671	24	2	1352
16	2	810	1751	28	2	1432
17	2	835	1831	7	2	1512
18	1	871	1903	4	2	1592
19	1	961	1999	13	3	1662
20	3	1010	2180	20	3	1732
21	1	1089	2252	27	3	1802
22	1	1108	2348	30	1	1874
23	4	1238	2576	29	1	1970
24	2	1238	2656	18	1	2066
25	1	1245	2728	19	1	2162
26	2	1262	2928	22	1	2258
27	3	1345	2998	21	1	2354
28	2	1365	3098	25	1	2450
29	1	1413	3170	23	4	2678
30	1	1519	3266	14	4	2768

FCFS *discipline*

Mean Delay = 922.9 sec
Max Delay = 1757.0 sec
z_{FCFS} = 53577.0 sec

CIH *algorithm*

Mean Delay = 605.3 sec
Max Delay = 2095.0 sec
z_{CIH} = 43552.0 sec

Lower bounds

LB1 = 38437.5 sec
LB2 = 22624.0 sec

Performance values

$100(z_{CIH}\text{-}LB1)/LB1 = 13.3\%$
$100(z_{FCFS}\text{-}z_{CIH})/z_{CIH} = 23\%$

Table 5. Computational results on a realistic ASP instance (n = 30, congested case)

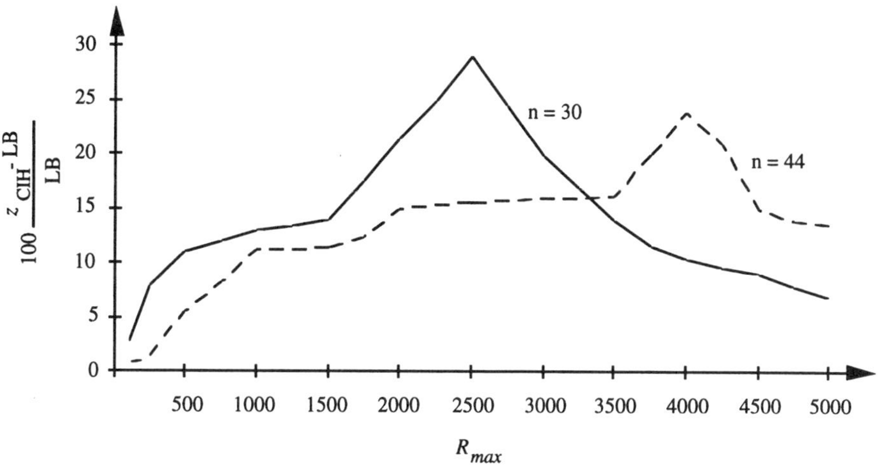

Figure 2. Duality gap between the CIH values and the best LB

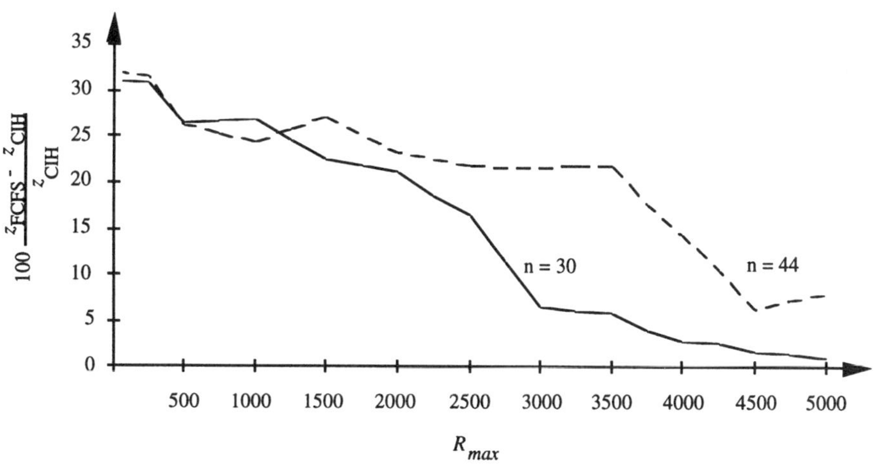

Figure 3. Comparing FCFS and CIH values

Moreover, in order to further investigate the behavior of the CIH algorithm, we simulated different series of problems for n = 30, 44 aircraft where the PLTs have been chosen uniformly distributed over the range [1, R_{max}] with R_{max} equals to 250, 500, 1000, 1500, 2000, 2500, 3000, 3500, 4000, 4500, 5000. Moreover, for each one of the 22 combinations of n, R_{max} we have considered 5 test problems and shown in Figures 2 and 3, respectively, the average values of the CIH algorithms compared to the FCFS solution and to the best lower bound LB = max[LB1, LB2].

The time spent by the CIH algorithm for all test problems was always less than 2 seconds.

7. A Dynamic Model for the ASP

In this section, we describe a sequencing model for the ASP in which we remove the assumptions d) and e) of Section 2, considering that the entry times of the aircraft are rather unknown in a operating environment. The scheduling operation is seen here as a dynamic task and the landing sequence is recomputed whenever the entry time of a future aircraft is known.

For this reason, the position of an individual aircraft, in rearranging the sequence, might be shifted backward many times producing an excessive delay. Moreover, if an aircraft gets close to the runway, its resequencing could be undesirable from the operators' point of view, because it would imply further changes in the maneuvering to be implemented while the aircraft is approaching the runway.

For taking into account the above observation we modify the model introducing two constraints parameters related to the number of shifting of an individual aircraft.

- **MPS:** Maximum Position Shifting. The MPS was defined by Dear [7]. For each aircraft the maximum number of shifts (either backward or forward) between the FCFS sequence and the actual landing sequence must be at most equal to the MPS parameter.

- **RPS:** Relative Position Shifting. The RPS represents the maximum number of shift operations *relative to the position* that an aircraft occupies in the sequence. This parameter can be defined arbitrarily by the operators for any subsequence of the actual landing sequence. For instance, aircraft in position 1, 2, 3, 4 (close to the final approach at the runway) can have RPS set to zero if we want their position to be frozen, while aircraft in the central part of the sequence can have RPS set to three, and those at the end RPS set to 6. Any other assignment of RPS is allowed from the model point of view.

The MPS is a well known parameter and it used to avoid aircraft to be excessively delayed. The RPS is related to new concept which should allow the operator to maintain some order when the sequence is rearranged. As an example, let us suppose that, on a sequence of 30 aircraft, the operator defines three subsequences [1–5],[6–19], [20–30] with RPS 0, 3, 5, respectively. In that case, the position of the first five aircraft [1–5] will be frozen (i.e., no changes will be produced by rearranging the sequence when a new entry time of on aircraft is known). Aircraft in position [8–19] could be shifted by three positions (if their MPS allows it) and those in positions [20–30] could be shifted by five position (if their MPS allows it).

Note that a shifting operation is considered feasible if both the MPS and RPS constraints are satisfied. Subsequences, RPS and MPS values can be arbitrarily defined by the controllers and we believe this makes the model more flexible in an operating dynamic environment.

Next is a description of an algorithm for the dynamic case, denoted CSH (Cheapest Shift Heuristic). The CSH algorithm is based on the following idea. Whenever a new entry time of an aircraft $j+1$ is known the corresponding PLT_{j+1} is computed and the aircraft is added at the end of the sequence. From this sequence S, which is a feasible one, CSH looks for a better solution among a set $X(S)$ of adjacent feasible solutions. Given a feasible solution S, of cost $f(S)$, the set $X(S)$ is defined as the set of feasible sequences obtained from S moving forward an aircraft. A more detailed description of the heuristic is as follows:

CSH Algorithm

Step 1. Let *best* := MAXREAL.

Step 2. While $X(S) \neq \Phi$ and $f(S) < best$, do

Step 2.1. Let $best := f(S)$;

Step 2.2. Let $S' \in X(S)$ such that $f(S') \leq f(S'')$ for each $S'' \in X(S)$;

Step 2.3. If $f(S') < best$,

let $S := S'$;

let $best := f(S')$.

The algorithm has been tested on the same test problems examined for the static case. In particular, we report the results for the case of the realistic test problem of 44 aircraft (see the previous section). We make the assumption that each aircraft is seen either 750 seconds or 1500 seconds before its PLT. With reference to the MPS and RPS parameters, we examine the operating scenarios generated as follows. We consider four different MPS values (2, 4, 6, 8). For each MPS value, we examine three different configuration, denoted as A, B, C, of subsequences and RPS values.

- Configuration A corresponds to consider the sequence divided in seven subsequences S_1 [1–4], S_2 [5–9], S_3 [10–14], S_4 [15–19], S_5 [20–24], S_6 [25–28] and S_7 [29–44]; we consider $RPS(S_1) = 0$, $RPS(S_2) = 1$, $RPS(S_3) = 2$, $RPS(S_4) = 3$, $RPS(S_5) = 4$, $RPS(S_6) = 5$ and $RPS(S_7) = 20$, respectively.

- Configuration B corresponds to consider the sequence divided in eight subsequences S_1 [1–3], S_2 [4–7], S_3 [8–10], S_4 [11–13], S_5 [14–15], S_6 [16–18], S_7 [19–21] and S_8 [22–44]; we consider $RPS(S_1) = 0$, $RPS(S_2) = 1$, $RPS(S_3) = 2$, $RPS(S_4) = 3$,

$RPS(S_5) = 4$, $RPS(S_6) = 5$, $RPS(S_7) = 6$ and $RPS(S_8) = 20$, respectively.

- Configuration C corresponds to consider the sequence divided in two subsequences S_1 [1–3] and S_2 [4–44]; we consider $RPS(S_1) = 0$ and $RPS(S_2) = 20$, respectively.

See Figure 4 for a synthetic visual representation of the actual constraints.

In Table 6 we have reported the results of CSH algorithm on the test problem of 44 aircraft: the first column shows how many seconds an aircraft is seen before its PLT; columns number two and three show the MPS and RPS configurations; finally, in the last two columns the maximum and mean aircraft delays are showed.

Comparing the mean and maximum delays showed in Table 6 with that ones occurred with the FCFS discipline for the same instance (see Table 3), it seem sthat a better sequencing could be obtained also for the dynamic case. Moreover, the time spent by a single run of the CSH algorithm was less than 0.5 seconds, and it seems fast enough to be implemented in an on-line advisory system.

8. Conclusions

In this paper, we have presented models and algorithms for the Aircraft Sequencing Problem where the objective is the minimization of the cumulative delay. Two models has been considered. The first one assumes that the entry of aircraft are all known, while the second consider the dynamic environment in which aircraft arrive continuously and entry times are known when a landing sequence have been already computed. For both models we presented fast heuristics and gave computational results on set of data which should be similar to real ones. Future research will be devoted for further testing of the proposed algorithms in different practical situations.

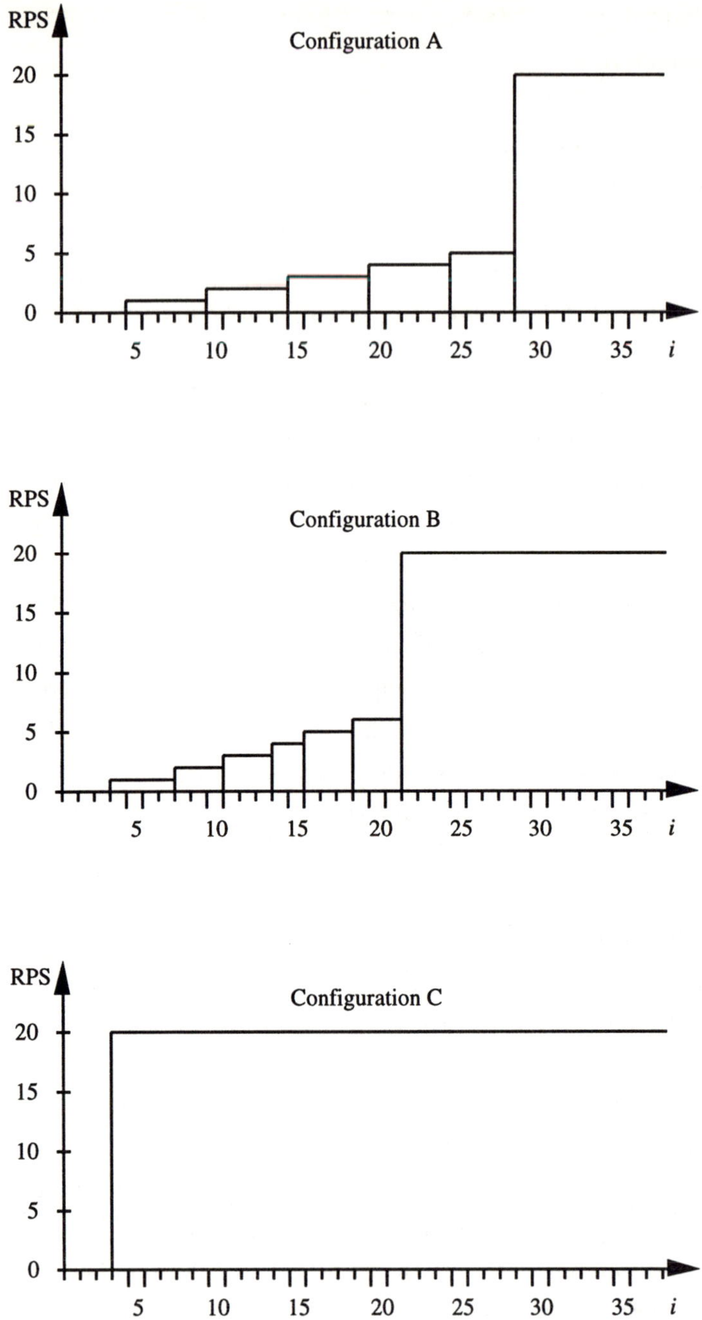

Figure 4. Examples of RPS assignments

ΔT (sec)	MPS	RPS	Max Delay (sec)	Mean Delay (sec)
750	2	A	1381.0	698.8
750	2	B	1381.0	698.1
750	2	C	1285.0	657.2
750	4	A	1443.0	590.2
750	4	B	1431.0	579.5
750	4	C	1527.0	630.7
750	6	A	1539.0	590.0
750	6	B	1149.0	564.4
750	6	C	1193.0	592.3
750	8	A	1539.0	590.0
750	8	B	1149.0	564.4
750	8	C	1193.0	592.3
1500	2	A	1285.0	654.0
1500	2	B	1285.0	654.0
1500	2	C	1285.0	654.0
1500	4	A	1431.0	579.5
1500	4	B	1315.0	663.9
1500	4	C	1527.0	630.7
1500	6	A	1149.0	572.9
1500	6	B	1245.0	602.8
1500	6	C	1149.0	564.4
1500	8	A	1149.0	572.9
1500	8	B	1363.0	595.5
1500	8	C	1159.0	564.4

Table 6. Computational results for the dynamic case (n = 44)

References

[1] G. Andreatta, G. Romanin-Jacur, Aircraft flow management under congestion, *Trans. Science*, 21, 249–253, 1987.

[2] L. Bianco and M. Bielli, System Aspects and Optimization Models in ATC Planning, in: L. Bianco, A. R. Odoni (Eds.), *Large Scale Computation and Information Processing in Air Traffic Control*, Springer Verlag, 47–99, 1993.

[3] L. Bianco, P. Dell'Olmo and G. Giordani, *One machine scheduling with ready times and sequence dependent processing times: preliminary results*, Report 408, IASI-CNR, Rome, 1995.

[4] L. Bianco, A. Mingozzi and S. Ricciardelli, The Traveling Salesman Problem with Cumulative Costs, *Networks*, 23, 81–91, 1993.

[5] L. Bianco, G. Rinaldi and A. Sassano, Scheduling tasks with sequence-dependent processing times, *Naval Res. Log.*, 35, 177–184, 1988.

[6] N. Christofides, A. Mingozzi and P. Toth, State space relaxation for the computation of bounds to routing problems, *Networks*, 11, 145–164, 1981.

[7] R. G. Dear, *The Dynamic Scheduling of Aircraft in the Near Terminal Area*, FLT R76.9, Flight Transportation Laboratory, M.I.T., Cambridge, 1976.

[8] M. Fischetti, G. Laporte and S. Martello, The delivery man problem and cumulative matroids, *Opns. Res.*, 41, 1055–1064, 1993.

[9] M. Held, P. Wolfe and H.P. Crowder, Validation of subgradient optimization, *Math. Prog.*, 6, 62–88, 1974.

[10] R.L. Graham, E.L. Lawler, J.K. Lenstra and A.H.G. Rinnooy Kan, Optimization and approximation in deterministic sequencing and scheduling theory: a survey, *Ann. Discrete Math.*, 5, 287–326, 1979.

[11] H. N. Psaraftis, *A Dynamic Programming Approach to the Aircraft Sequencing Problem*, FTL R78-4, Flight Transportation Laboratory, M.I.T., Cambridge, 1978.

[12] H. N. Psaraftis, A dynamic Programming Approach for Sequencing Identical Groups of Jobs, *Opns. Res.*, 28, 1347–1359, 1980.

[13] J.-C. Picard and M. Queyranne, The time-dependent traveling salesman problem and its application to the tardiness problem in one-machine scheduling, *Opns. Res.*, 26, 86–110, 1978.

[14] A.H.G. Rinnooy Kan, *Machine scheduling problems: classification, complexity and computations*, Martinus Nijhoff, The Hague, Netherlands, 1976.

[15] W.E. Schrage, A proof of the optimality of the shortest remaining processing time discipline, *Opns. Res.*, 16, 687–690, 1968.

[16] C.S. Venkatakrishnan, A. Barnett and A.M. Odoni, Landings at Logan Airport: Describing and Increasing Airport Capacity, *Transp. Sci.*, 27, 211–227, 1993.

HUMAN CENTERED DEVELOPMENT OF INFORMATION SYSTEMS AND DECISION AIDS IN ADVANCED AIR TRAFFIC MANAGEMENT SYSTEMS

R. John Hansman

Professor of Aeronautics & Astronautics
Head of the Division of Humans & Automation
Director of the International Center for Air Transportation
Director of the Aeronautical Systems Laboratory
Massachusetts Institute of Technology
Cambridge, Massachusetts, USA

James K. Kuchar

Assistant Professor of Aeronautics & Astronautics
Massachusetts Institute of Technology
Cambridge, Massachusetts, USA

Eric. N. Johnson

Research Assistant, Dept. of Aeronautics & Astronautics
Massachusetts Institute of Technology
Cambridge, Massachusetts, USA

Human performances are central in Advanced Air Traffic Management systems. Information systems and decision aids in these advanced systems will be simultaneously driven by technical and human capabilities coupled with operational requirements. Current ATM information technologies are reviewed and issues of human operation of automated systems are discussed. An Integrated Human Centered Systems Approach which considers the human controller as a functional component of the closed loop information system, is suggested.
To effectively apply it to ATM information systems, simulation and evaluation tools are developed.

1. Introduction

The development of new information sub-systems within Air Traffic Control architectures has traditionally been driven by improvements in surveillance and communications technologies coupled with operational and traffic requirements. The human controller has always been recognized as the key component in the Air Traffic Management (ATM) system. In the past, the human controller was expected to compensate for limitations in technology. The outstanding performance of modern Air Traffic Management Systems is a testament to the skill, training, and adaptability of Air Traffic Controllers.

The explosive increase in information and communication technologies has changed the process and requirements for the development of human-automation interfaces. In future information intensive environments such as advanced ATM systems it is likely that the capabilities of the human controller will become the limiting factor in the information system. While the bandwidth of the information technologies will increase significantly, the "bandwidth" of the human controller will remain relatively constant.

In order to improve the performance of the human controller in advanced ATM systems, a variety of information systems and decision aiding technologies will be proposed. Unless careful consideration of the human controller is given in the development process, the overall system performance may actually degrade due to factors such as information saturation, divided attention, and task load association with management of the decision aids. This paper discusses an integrated "human centered" systems development process where the human is considered as a functional component of the closed loop ATM system.

2. ATM Systems Development Opportunities

A variety of technical and operational factors have combined to provide a rich environment for the development of advanced ATM systems. Operational factors include: high peak traffic flows resulting in terminal area, enroute and oceanic saturation and delays. These high peak traffic levels have resulted in pressure to operate at reduced separation and to achieve higher level of control accuracy to maximize the utilization of limited resources such as runway occupancy time.

Technical factors include new communications, navigation, surveillance, decision aid and flight management systems. Several of the more important technologies are discussed briefly below.

Datalink - A variety of datalink systems are in operation and in development. These range from VHF-ACARS with limited range and bandwidth to Mode S with a higher bandwidth but also limited range to satellite datalinks which have high bandwidth and can achieve continual communication over oceanic ranges. The datalink offers the potential for direct communication between ATM ground processors and aircraft Flight Management Computers. This capability is already in use for Pre-Departure Clearance delivery through VHF-ACARS for some aircraft. There is some concern that managing the datalink communications may result in an increase in controller workload.

Computer Assisted Hand-offs - Datalink communications between controller workstations has been used to improve efficiency by transmitting preview information and automating parts of the hand-off process.

TCAS - The Traffic Alert and Collision Avoidance System which is equipped on many air carrier aircraft has altered the dynamics of the controller-pilot relationship. With the cockpit display of traffic (i.e. distributed surveillance) enabled by TCAS, there is performance and social pressure for the authority and responsibility for traffic separation to decentralize.

GPS - The use of Global Navigation and Surveillance Systems such as GPS for worldwide area navigation represents a significant change in navigation capability and route structure architecture. As radio navigation waypoints become less important, the route structure can more closely reflect the user needs rather than the geographical location of the navigation beacons.

ADS - Area navigation systems coupled with data downlink of position and altitude allow surveillance of regions which are inaccessible by radar. Examples include oceanic regions, terrain masked regions, and regions with insufficient traffic density to warrant radar coverage.

FMS - Advanced aircraft Flight Management Systems have the capability to fly highly accurate, complex trajectories in space and time. In addition, these systems have the capability of planning fuel efficient decent profiles. One problem in the development of FMS ATM procedures is that not all aircraft are FMS equipped and there are significant variations in FMS capability even between similar aircraft due to manufacturer and software version differences.

Automated Alerting Systems - A low level of decision aiding in ATM systems are automated alerting systems such as the Minimum Safe Altitude Warning (MSAW) or the Conflict Alert (CA).

Decision Aid and Optimization Technologies - A variety of tactical and strategic decision aids are under development for ATM systems with the most mature probably being CTAS. In addition, many algorithmic techniques have been developed.

3. Human Issues in ATM Automation

The evolution of automated ATM systems from simple alerting systems to more complex decision aids such as schedulers and descent advisors has paralleled the evolution of automation in other supervisory control systems such as aircraft flight management systems. Many of the generic human-automation issues which emerged in the aircraft domain [1-3] are also applicable in the ATM domain. In addition there are issues specific to ATM operations. A few of these are discussed briefly below.

System Performance - One basic tenet of automation systems is that the automation must improve the performance of the coupled human/system. While this seems obvious it implies that the test for applicability of any automation system is to compare it against the performance of a non-automated alternative. These comparisons must include the full range of potential operating conditions including non-normal situations as well as partial and complete automation system failure.

Situational Awareness - One of the critical factors in ATM operations is for the controller to have a sufficient understanding (i.e. picture) of all relevant factors to allow well informed control decisions and actions. Currently the term Situational Awareness is used to describe this meta-level understanding [4,5]. As the control environment becomes more complex, crowded, unstructured (i.e. Free Flight), and constrained, it will be more difficult to maintain Controller Situational Awareness.

Attention Limitations - It is well known that humans are poor monitoring agents and monitoring performance will degrade with boredom and loss of attention [6-7]. If automaton is poorly applied, it is possible for the human to be allocated a monitoring task which is out of the primary control loop. This can result in degraded performance. In addition, the controller may be expected to suddenly resolve situations beyond the capabilities of the automation. If the controller is not actively involved in the decision loop, he or she will require a finite time to orient and develop sufficient Situational Awareness to provide appropriate control actions.

Information Overload - One of the key human performance issues in advanced information systems is to determine the appropriate quantity, format, and pre-processing of information to provide to the operator to prevent loss of situational awareness due to information overload. Essentially,if the bandwidth of the important and available information is greater than the human's capacity, then information will be missed. For multi-tasking environments such as ATM, this problem is exacerbated by a loss of multi-tasking capability as information saturation is approached. Unskilled controllers will tend to focus on single information elements while more skilled operators can prioritize information to effectively reduce the incoming bandwidth.

Understanding of the Automation Criteria - In an effort to minimize information overload, automation is often used to pre-filter and condition the information which is presented to the controller. In addition, the decision aids may make recommendations based on some optimization criteria. These decision aids and displays can improve performance if the information is presented in an intuitive format and the optimization criteria is clear to the controller. However, performance can be degraded if the controller does not clearly understand the roles and criteria of the automation. This has been a problem in some complex aircraft Flight Management Systems where the underlying automation structure is not clear to the user [3].

Changes in Communications Modes - One impact of datalink technologies is the potential change in the communication of control instructions. Many of the existing flight procedures and route structures have been developed and named in order to allow efficient voice communication over VHF and HF links. With the ability to digitally communicate a complex series of 4D waypoints (Latitude, Longitude, Altitude, Time) the potential exists for the use of unnamed procedures which cannot be communicated by voice in the event of a datalink failure.

Loss of Party Line Information - One of the major concerns of the extensive use of datalink communications is the loss of "Party Line" information and the Situational Awareness which is achieved by monitoring secondary voice conversations on VHF and HF channels [8,9]. Based on survey and simulator studies, the important areas appear to be traffic and weather information. Potential "Party Line" compensation mechanisms are under study including TCAS and datalink distribution of weather information.

Unexpected Compensatory Behavior - In some cases, technology can have unexpected and undesirable side effects. For example, the presence of TCAS reduced terminal area capacity in some cases because some controllers added additional “in trail” separation to TCAS aircraft to avoid inadvertent violation of in-trail wake vortex criteria which could be observed and reported by pilots.

Human Acceptance of Automation - There are many social and psychological factors which influence how automation will be accepted. In systems like ATM where the human is ultimately responsible, the controller must develop “trust” in the automation [7]. While there are many factors which influence “trust”, it is clear that the automation must be reliable and the controller must have a clear understanding of the operation and limitations of the automation. A second factor in the acceptance of automation is any perception of “threat” that the automation poses for the controller. This “threat” can be minimized if the automation is seen as a vehicle to enhance the controller capabilities rather than to supplement them. One mechanism for improving the acceptability of automation is to include current controllers early in the development process.

Human Reliance on Automation - As automation is used to improve the performance of the human/automation system, issues of human reliance and reliability must be considered. If the potential exists for automation failure, procedures and policies must exist to maintain controller base skill levels for adequate operation in the non-automated state. If automation is critical to the task then issues of fault tolerance and system redundancy must be addressed.

4. Integrated Human Centered Systems Approach

The Integrated Human Centered Approach is well suited to the development of advanced ATM information systems. This approach applies known techniques of human centered design but maintains a Systems Engineering method to the development process. Within this context, the human is considered as a functional component of the closed loop information system. System level trades are considered to evaluate the allocation of capability between the human and other components of the information systems such as the sensors, displays, or automation systems. A key element of the integrated approach is practical consideration for the actual operating environment. Many proposed information system elements which look good on paper fail in operation. A simple example would be a decision aid which did not consider "real world" behavior such as the variability in pilot response time to controller instructions or the possibility of a blocked communication. The approach can be applied to one or more functional paths (i.e. tasks) which the human-ATM system is expected to accomplish. The key elements of the approach are outlined below.

Identify Information Requirements and Issues - This is accomplished using a variety of methods depending on the information system under study. Typically a functional analysis and time line of the operation is conducted and a base set of information requirements is developed. For evolutionary systems, it is common to conduct focused interviews and surveys of controllers and flight crews currently operating similar systems or conducting similar operations to identify key issues and obtain operational insight. This can be particularly beneficial in that by engaging the controller community early in the development process, the community has "ownership" of the approach and is less likely to be threatened by the automation.

Development of Prototype Systems and Options - Based on the results of the Information Analysis and an assessment of technological capability and readiness, several prototype information systems are typically developed to explore various system options or to address issues raised in the Information Analysis. These systems are typically developed on rapid prototyping part-task simulators based on graphical workstations and virtual reality technologies which allow easy exploration of different system options. In many cases, fundamental issues are identified and resolved in the prototyping process. The degree of fidelity is matched to the functional requirements of the task. For many information systems, the issues are cognitive and relatively low fidelity simulation is sufficient for preliminary studies. An example of a typical simulation facility is the MIT Advanced Cockpit Simulation Facility shown in Fig. 1.

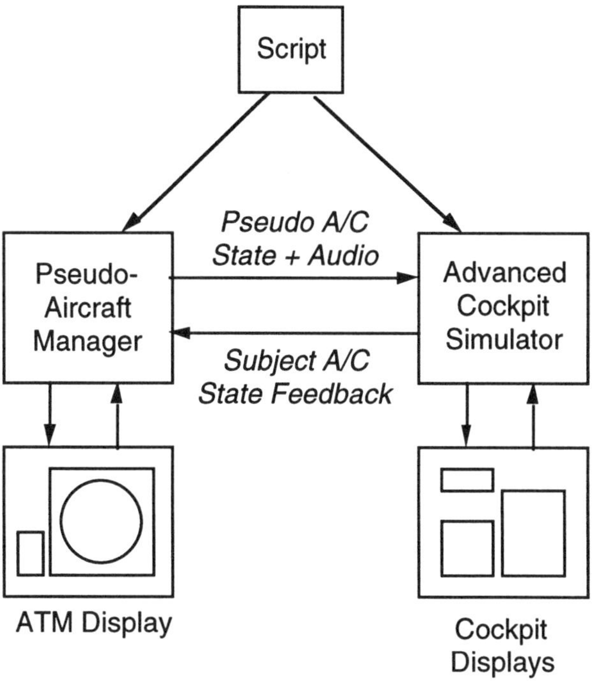

Figure 1. MIT Simulation Facility

Simulation Evaluations - Simulation evaluations of prototype information system options are conducted using controller subject populations. Both performance metrics and subjective evaluations are used for this purpose. Performance metrics include factors such as: delivery accuracy, operational errors, and situational awareness. Experienced controllers rarely make operational errors however it is possible to develop challenging probe scenarios which test the level of controller situational awareness. Traditionally, the simulations at this level involve pseudo-pilots who are difficult to script into repeatable scenarios. The automated generation of controlled multi-agent scenarios using workstation technology is described in a section below. Based on the results of the preliminary evaluations, the information requirements and prototype systems may be modified and further simulations would be conducted in an iterative process.

Integrated Simulation Testing - For some development systems, it is necessary to run more complex simulation studies to investigate the interaction dynamics between multiple agents (controllers and pilots). This can be conducted in distributed simulation facilities with combinations of real and simulated systems.

System Evaluation - Based on the result of the simulation evaluations, system level assessments are conducted with regard to the potential impact of the information system. This would include development requirements, system effectiveness, safety implications and cost-benefit analysis.

Field Development Phase - For information systems which have favorable cost-benefit profiles, preliminary systems are developed for field studies with live controllers and ultimately live aircraft. The results of these field studies are used to develop system specifications which are used to procure operational systems.

5. Tools for Human Centered Development of ATM Information Systems

In order to effectively apply the human centered development approach to ATM Information Systems, several simulation and evaluation tools have recently been developed within the MIT Aeronautical System Laboratory. Several of these are discussed briefly below.

Robust Multi-Agent Situation Generation - In order to probe controllers' situational awareness and to comparatively test different display/information system options in simulation experiments it is often necessary to have carefully controlled and repeatable situations occur. Traditionally, ATM simulations at this level have relied on human pseudo-pilots who are difficult to script into repeatable scenarios. If a precise situation is required, such as a collision threat, a small shift in timing can significantly alter the perceived threat to the controller. The variability in voice responses by pseudo-pilots is also an issue.

An automated technique for the generation and action of multiple agents (i.e. targets) in ATM simulations has been developed and is shown schematically in Fig. 2 [10].

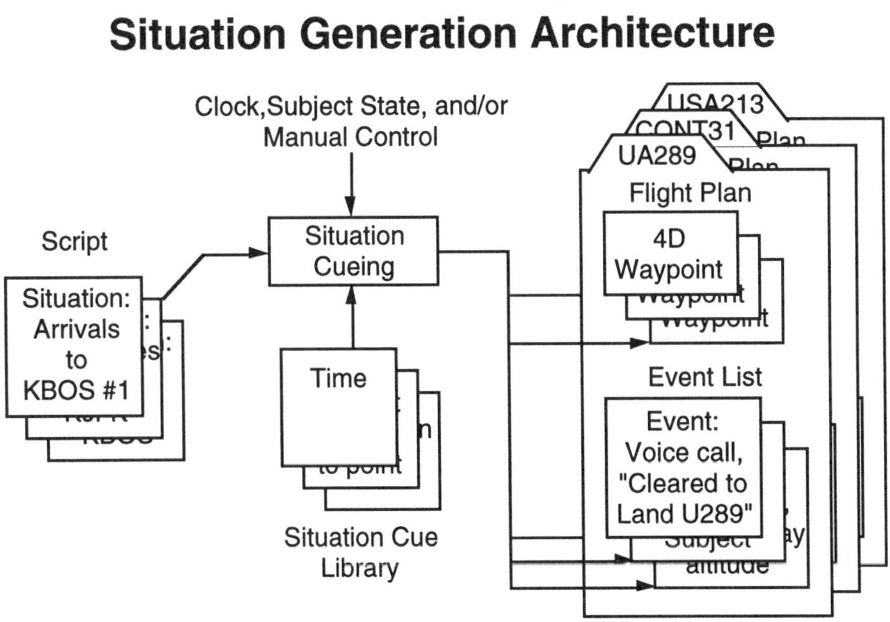

Figure 2. Robust Multi-Agent Situation Architecture

To produce robust situations, a framework has been developed and implemented using closed-loop feedback. A pre-defined script determines the situations that are to occur in the experiment. During the experimental run, the controller's actions and key agent states are continually fed back to the situation generator, which steers and times the other agents actions in order to produce the desired situations. The scripts include vehicle trajectories and voice responses which are digitally pre-recorded. Several techniques are used to trigger and control events within the scenarios. These include the use of 4D waypoints, the use of target relative waypoints, controller action cueing and experimenter cueing. This system has been used for several experiments testing the impact of TCAS on Party Line Information use and traffic displays for closely-spaced parallel runway operations.

Probabilistic Analysis of Hazard Situations and Alerting Thresholds - One of the most persistent issues in the development of automated decision aids is the establishment of alerting thresholds. There is a classical trade between the Probability of Correct Detection versus the Probability of False Alarm. This issue is expected to be even more critical in unstructured environments such as "Free Flight" where the dependence on hazard alerting will be greater and the alerting criteria will be more complex.

A probabilistic analysis methodology has been developed to examine the fundamental relationships between the parameters affecting alerting system design (such as sensor accuracy or alert threshold method). The methodology is created through an investigation of the basic issues relating measurement accuracy, expected aircraft performance, hazard severity and extent, and situation geometry. The methodology, shown schematically in Fig. 3, can be used to evaluate alerting systems both in terms of achievable false alarm rate and by highlighting areas where further improvements in system design may be of value. The methodology can also be used to parametrically or statistically investigate the effect of different assumptions of human response to an alert such as delay latency and aggressiveness of action.

The tradeoff between False Alarms and Correct Detections can be shown using a System Operating Characteristic (SOC) curve which is similar to the Receiver Operating Characteristic curve of signal detection theory [11].

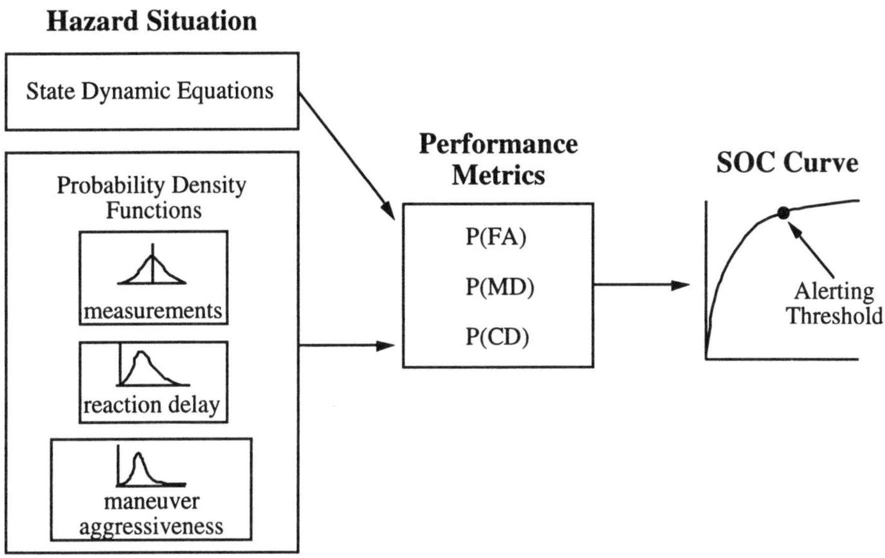

Figure 3. Probabilistic Analysis Methodology

The methodologies been applied to a variety of alerting systems including Ground Proximity Warning Systems (GPWS) and the Traffic Alert and Collision Avoidance System (TCAS). In the TCAS example, a situation was analyzed in which an intruding aircraft is descending toward the subject aircraft but could level off safely with some probability. The tradeoff between False Alarms and Correct Detections is shown on the System Operating Characteristic (SOC) curve in Fig. 4. As the SOC curve shows, the updated TCAS Version 6.04A threshold is located such that it produces a high probability of a Correct Detection while minimizing the probability of a False Alarm.

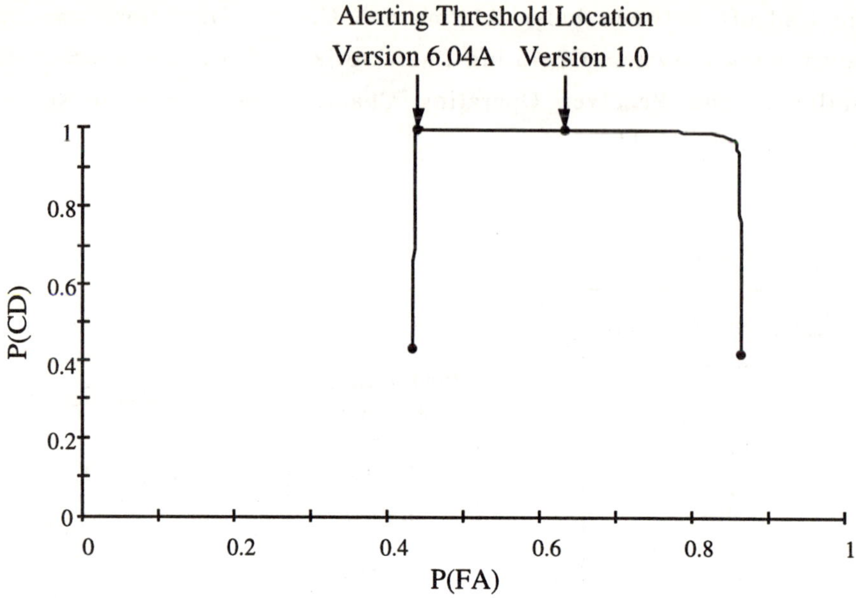

Figure 4. TCAS System Operating Characteristic Curve

6. Conclusions

Human performance considerations are expected to be central to the performance of Advanced Air Traffic Management Systems. The development of information systems and decision aids in these advanced systems will be simultaneously driven by technical and human capabilities coupled with operational requirements. An Integrated Human Centered systems approach is suggested which considers the human controller as a functional component of the closed loop information system.

Acknowledgments

The authors have been supported by the National Aeronautics and Space Administration and the Federal Aviation Administration.

References

[1] E.L. Weiner and R.E. Curry. Flight deck automation: promises and problems. *Ergonomics*, 23:995-1011, 1980.

[2] N.B. Sarter and D.D. Woods. *Autonomy , authority, and observability: properties of advanced automation and their impact on human-machine coordination*. Proceedings of the 6th IFAC/IFIP/IFORS/IEA Symposium on Analysis, Design and Evaluation of Man-Machine Systems, Cambridge, MA, 1995.

[3] S. Vakil, A. Midkiff, T. Vaneck and R.J. Hansman. *Mode awareness in advanced autoflight systems*. Proceedings of the 6th IFAC/IFIP/IFORS/IEA Symposium on Analysis, Design and Evaluation of Man-Machine Systems, Cambridge, MA, 1995.

[4] M.R. Endsley and E.O. Kiris. The out-of-the-loop performance problem and level of control in automation. *Human Factors*, 37(2):381-394, 1995.

[5] M.R. Endsley and M.D. Rodgers. *Situation awareness information requirements for en route air traffic control*. Report DOT/FAA/ AM-94/27, 1994.

[6] E.L. Weiner and D.C. Nagel. *Human factors in aviation*. Academic Press, 1988.

[7] T.B. Sheridan. *Telerobotics, automation, and human supervisory control*. The MIT Press, 1992.

[8] A.H. Midkiff and R. J. Hansman. Identification of important "party line" information elements and implications for situational awareness in the datalink environment. *Air Traffic Control Quarterly*, 1(1): 5-30, 1993.

[9] A. Pritchett and R.J. Hansman. *Preliminary analysis of pilot rankings of "party line" information importance*. Seventh International Symposium on Aviation Psychology, April 1993.

[10] E. Johnson and R.J. Hansman. *Multi-agent flight simulation with robust situation generation*. Report ASL-95-2, Cambridge, MA 1995.

[11] J. Kuchar and R.J. Hansman. *A unified methodology for the evaluation of hazard alerting systems*. Report ASL-95-1, Cambridge, MA, 1995.

SAFELY REDUCING DELAYS DUE TO ADVERSE TERMINAL WEATHER*

James E. Evans

Leader, Weather Sensing Group
Massachusetts Institute of Technology
Lincoln Laboratory
Lexington, Massachusetts 02173-9108, USA

This paper focuses on assessing how much of the $ 3B (US) per year in adverse terminal weather delays is "avoidable" and which Air Traffic Management (ATM) investment options would be most effective in eliminating the delay. We propose that such delay can be modeled as a combination of delays due to two mechanisms: one which is "linear" in terms of the impact of key variables and the other is a nonlinear queuing model. We then consider various options for reducing the delay in the context of these models, with the prime objective being to point out where the greatest opportunities exist for delay reduction. We also make suggestions for better assessing the potential benefits of the various ATM options.

1. Introduction

Adverse terminal weather seriously impacts the aviation system users in the form of delays, cancellations, diversions, increased controller workload, reduced scheduling integrity and loss of payload as well as being a major cause of accidents. All of these have substantial economic impacts. To illustrate the magnitude of the numbers, the estimated cost of delays to the U.S. air carrier system is $ 5B per year, of which 65 percent is generally attributed to weather (as communicated by Mr. L. Kiernan of the FAA National Planning Division to the FAA National Capacity Indicator Forum, 1 Sept. 1994).

---

* This work was sponsored by the Federal Aviation Administration. The views expressed are those of the author and do not reflect the official policy or position of the U.S. Govenment.

If a substantial fraction of this delay could be shown to be reduced by appropriate Air Traffic Management (ATM) methods, it would clearly motivate very substantial investments in ATM research and development.

Hence, there has been considerable interest in assessing how much of this delay is "avoidable" and which ATM investment areas would be most fruitful in eliminating the delay. We propose two conceptual models for delay and discuss key elements of the U.S. aviation system that are germane to these models (corresponding studies in Europe and Asia would be interesting).

Next we consider various options for reducing the delay in the context of these models, with the prime objective being to point out where the greatest leverage needs to be, along with suggestions for better assessing the potential benefits of the various ATM options.

2. Background Information

Table 1 [Weber, et al., 1991], shows the results of extrapolating delay data for Chicago's O'Hare Airport during various weather events to a number of airports based on the differences in traffic into the airport. A subsequent comparision of delays per aircraft at O'Hare with delays at Minneapolis and Dallas airports [Evans, 1995] suggested that the O'Hare results were not atypical.

We see that there are major differences between the various airports in terms of the types of weather phenomena which lead to delays. The key question is the extent to which the delay that occurs is in fact avoidable.

Thunderstorms account for approximately 50 percent of the delay in table 1. A common misconception is that thunderstorm delay is unavoidable because the airport is closed due to hazardous weather. Detailed studies of thunderstorm delay in the terminal area by our group have shown that the bulk of the delay arises from difficulties in:

AIRPORT	DAILY OPS	CLIMATOLOGY (Days per Year) T-Storm	Hvy Fog	Lo Vis.	DELAYS > 15 Min. ANNUAL DELAY MIN X 1000 Th.	Hvy Fog	Lo Vis.	Wx.
Chicago	2175	38	16	109	412	94	185	87%
Atlanta	2156	50	30	136	538	174	229	90%
Los Angeles	1589	3	44	121	24	188	150	83%
Dallas	1578	45	11	86	354	47	106	87%
Denver	1438	41	10	57	294	39	64	85%
San Francisco	1255	2	17	101	13	57	99	74%
St. Louis	1178	45	11	156	265	35	143	89%
Boston	1162	19	23	125	110	72	113	84%
Phoenix	1142	23	2	5	131	6	4	72%
Detroit	1137	33	22	121	187	67	107	87%

Table 1. Weather Impact at U.S. Airports [from Weber, 1991]

1. Planning terminal routes when planes must be vectored around storms,

2. Anticipating when runways will be usable,

3. Anticipating the opening and closing of the "gates" into and out of the terminal area, and

4. Matching the rates of flow into the terminal with the effective airport capacity as opposed to outright airport closure.

Another principal cause of delay is the reduction in capacity with low ceiling and visibility (C&V) conditions at airports which have closely spaced runways or other operational restrictions during instrument meteorological conditions (IMC). As in the case of thunderstorms, we are not referring here to very low visibility conditions such that landing at all is a problem. Rather, one typically cannot use the runways as efficiently as is done during visual meteorological conditions (VMC). Table 2 compares the scheduled arrivals at various airports with typical effective capacities during IMC conditions on the basis of worst-case arrivals in a one-hour period and on the basis of average arrivals between 8 AM and 9 PM. Note that the effective IFR capacity for an airport can change significantly with different wind directions and speeds. The capacities shown in table 2 were typical restrictions imposed during the summer of 1994. Clearly at a number of airports, the onset of IMC conditions must surely lead to either delays and/or cancellations and diversions.

Figure 1 shows a very simple conceptual model for the aviation system focusing on key elements in controlling flows in the system. The role of the terminal and enroute controllers is well understood. It should be emphasized that controllers typically are concerned about handling the aircraft that are currently in their area of responsibility and hence have little concern for the forecast weather or traffic flow management.

Highest Worst-Case Hourly Deficit		Highest Average Hourly Deficit	
1. Kennedy	42	1. Boston	16
2. Boston	36	2. Newark	10
3. O'Hare	35	3. Kennedy	5
4. Dallas	34	4. St. Louis	4
5. Philadelphia	27	5. LaGuardia	3
6. Newark	23	6. O'Hare	0
7. Detroit	23		
8. St. Louis	22		

Table 2. Capacity Deficit During IMC Weather Relative to Arrival Demand for Various Airports

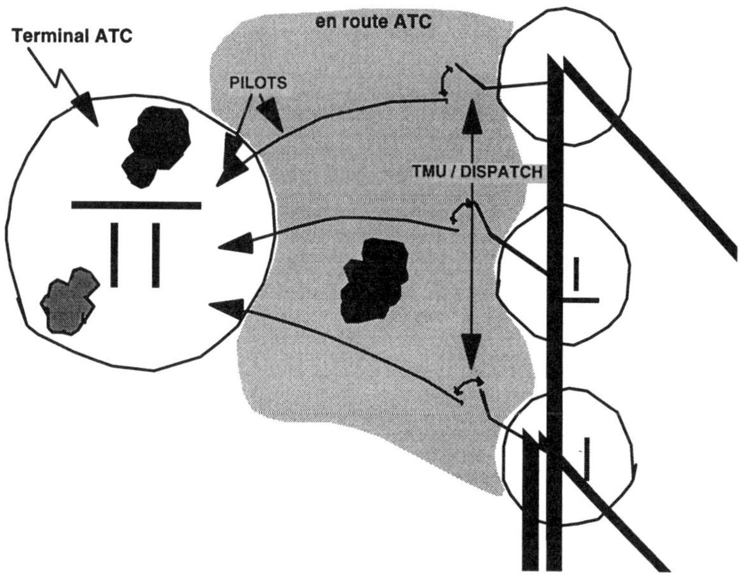

Figure 1. Effective terminal capacity factors

However, the role of traffic management unit (TMU) personnel in attempting to balance the air traffic demand with changes in the system capacity due to weather is not as well appreciated. In the U.S., there are TMU personnel at the national center ("central flow") as well as at each enroute center, and also at major terminals. When weather affects a particular terminal, the terminal and adjacent enroute center TMU personnel typically will attempt to compensate by slowing traffic into the terminal area and perhaps holding close-in aircraft. Longer lasting problems (especially those at major airports) may require "national" programs to delay a large fraction of the aircraft into a given airport, typically by imposing ground holds.

This planning role means that the TMU personnel must continually be anticipating weather impacts on operations rates throughout the aviation system. They are assisted in this task by a large real-time database which provides estimates of the expected traffic load and capacity in all sectors and key airports for each 15-minute period throughout a day, based on facility estimates, flight plans, plane locations, and winds [Jesuroga, 1993a and 1993b].

One of the major problems in understanding delay causality from recorded delay statistics is the "delay ripple" effect which arises when an aircraft is delayed on one leg of a flight (e.g., due to adverse weather) such that the next leg (and subsequent legs) flown by that aircraft that day also are delayed. In cases where the subsequent leg(s) are not weather impacted, the delay on the subsequent legs may not be attributed to terminal weather. DeArmon states that "delay ripple is in general pretty strong" and persists over a number of successive legs [DeArmon, 1992]. Hartman cites a case where the number of passengers delayed (down line impact) due to delay ripple was 27 times greater than the initial number delayed [Hartman, 1993].

3. Models for Delay

Delays can generally be modeled as either a fixed delay (typically to a finite number of aircraft) and/or a variable queuing delay wherein the effective capacity of the airport (or enroute sector) is less than the demand for a period of time. Since these ideas are pivotal to our overall discussion, let us consider them each in some detail.

3.1 "Linear" Delay Reduction

The first situation can arise with a transient event (e.g., a group of aircraft must fly a longer route) where there is no reduction in the overall average rate of aircraft movement. Figure 2 illustrates this for the case of a thunderstorm impacting an entry gate into a terminal area. Other examples of this include altitude changes to avoid clear air turbulence, avoiding missed approaches due to worse visibility conditions than expected, extra taxiing on the airport surface due to a runway change at a lightly loaded airport, and the use of inefficient descent trajectories due to air traffic procedures. A key element of this type of delay is that the benefit for improved performance is typically linear in each of the pertinent variables (e.g., traffic density, likelihood of occurrence, ability to realize the benefit in a given situation with an aviation system feature).

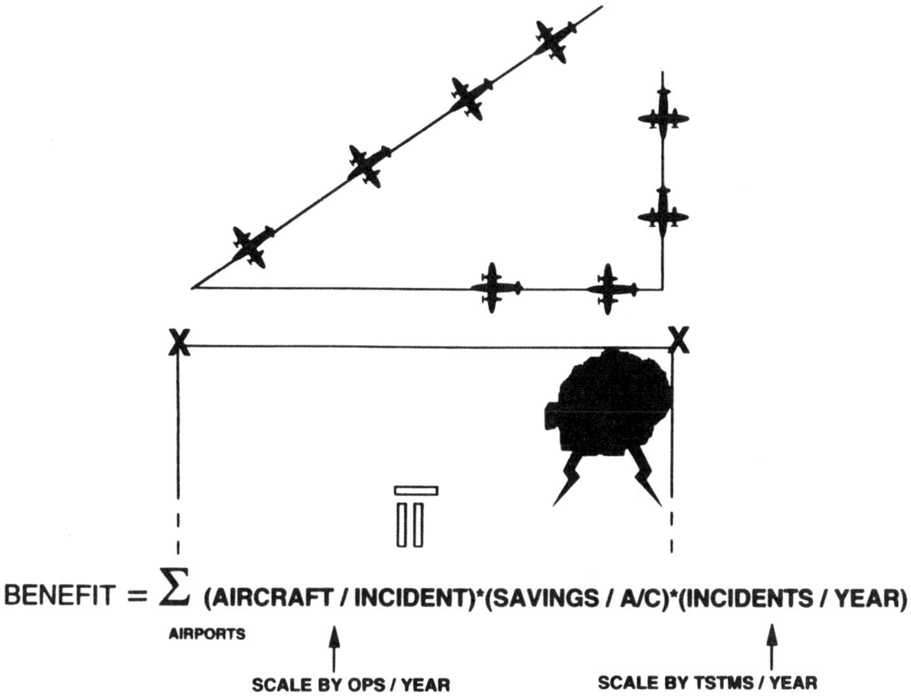

Figure 2. "Fixed" delay model

3.2 "Nonlinear" Delay Reduction

Figure 3 shows a simple example of the classic queuing situation where the weather reduces the effective capacity of an airport for some finite time. This simple queuing model can be used to address both air traffic control/airport reductions in effective terminal capacity <u>and</u> traffic flow management actions by interpreting:

1. The effective capacity as the minimum of the air traffic control/airport constraints on the traffic flow <u>and</u> the flow rate imposed by the TMU, and

2. The effective duration as the sum of the actual weather event duration <u>and</u> the time period over which an insufficient number of aircraft are available to land due to traffic management holds.

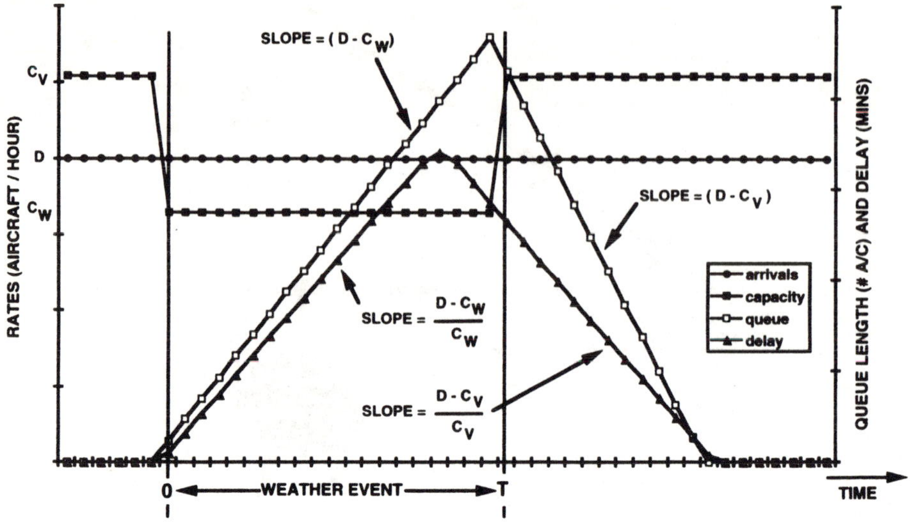

Figure 3. Queuing model for delay.
Example of Weather-Related Reduction in Terminal Capacity

Thus, for example, if an actual weather event lasts for two hours and creates a situation in which a number of aircraft desiring to land at the airport are held on the ground at the respective departure airports, the delay event may be viewed as continuing until the ground hold aircraft are released and land at the destination airport. If the minimum flight time for the aircraft being held on the ground is one hour, then the effective duration is at least three hours. The use of holding patterns near the airport (as in the FAA's Managed Arrival Reservoir technique) will result in a more complicated relationship than illustrated in figure 3, but the general principle still remains that ground holds increase the effective duration of a weather event.

It is straightforward to show that the accumulated delay for all the aircraft involved in the incident shown in figure 3 is

$$\Sigma \text{ delays} = 0.5\ T^2\ (D - C_W)\ (C_V - C_W)/(C_V - D) \qquad (1)$$

where:

C_W= capacity during adverse weather   D = demand
C_V= capacity during VMC weather   T = effective event duration

The dependence of delays on the traffic density and traffic flow management procedures here is quite nonlinear. For example, we see that small increases in the effective capacity during a weather event, C_W, can produce larger proportional reductions in the accumulated delay because C_W appears both in differences (e.g., a small increase in C_W will result in a larger fractional change in each of the differences) and in the product of terms.

Since T is squared, reducing the effective duration of a weather event (e.g., by better weather predictions and TMU decision making) can also produce large delay reductions. For example, if a good short-term prediction enables the TMU to reduce a 3-hour effective duration weather event to 2.5 hours, the accumulated delay is reduced 31 percent.

We have found that the calculations of the delay reduction using a refinement of the simple queuing model to include time varying demands and/or effective capacity can be carried out using a common personal computer spreadsheet. This approach is very computationally efficient and compares well with measured delays for the very limited number of cases analyzed to date.

3.3 Accounting for "Downstream" Effects

Historically, the FAA has assessed "downstream" effects by simulating individual flights in the overall aviation system with the NASPAC model [Frolow, 1989]. The advantage of such a simulation is that one could address at the same time downstream effects, including interactions between connecting flights and options that an airline might take to reduce delay impact on systems operations. However, the current models of this type are expensive to run and do not include many of the most interesting interactions (e.g., between connecting flights).

Consequently, several simple models have been used to account for downstream effects. One of these is a closed form answer obtained by assuming that aircraft can make up a fixed amount of delay (e.g., 20 minutes) per leg flown. Boswell has developed a model in which the amount of delay made up per leg is a random variable and has used this to determine a multiplier of approximately 1.8 for initial weather delays to arrive at a total weather delay in an Integrated Terminal Weather System (ITWS) benefits study [Boswell, 1997].

3.4 Application of Delay Models to Practical System Assessment

A large study of terminal operations efficiency benefits has been carried out as a part of the ITWS development program. In this study, estimates were made for situations involving both fixed delay reductions and queuing delay reduction by obtaining quantitative estimates from air traffic personnel (supervisors and TMU) and airlines for some 28 different benefits elements. The queuing model studies focused on increases in the effective arrival rates in the terminal area when thunderstorms impact the airport and on reducing the effective duration of low C&V events by short-term predictions for the end of the event.

In both cases, the qualitative behavior suggested by equation (1) occurred:

1. Relatively small changes in C_W (specifically 10 percent) could produce much larger proportional reductions in accumulated delay (e.g., 30-40 percent), and

2. Modest reduction in weather event effective duration (e.g., from 3 hours to 2.5 hours) by predicting the event time would produce large changes in delay

We have subsequently realized that in the ITWS study, a number of situations in which a fixed delay model was used (e.g., the benefit of anticipating a runway shift) should have been addressed using the queuing model since there will be a queue that forms while the runway shift is being accomplished that does not instantly disappear when the shift has been accomplished. This highlights the need to carefully analyze the aviation operations to decide on the best model to use before asking the questions that will fill in the model details.

4. Options for Reducing Delays

Three key elements in assessing options for reducing delays due to adverse terminal weather are the types of weather of concern, the airports of concern, and the philosophy used for scheduling flights.

Let us first consider the flight scheduling paradigm. In the U.S., flight scheduling is typically based on the assumption of fair weather and VMC procedures, with the understanding that delays may/will arise in adverse weather. Conversely, one could agree to forgo the additional capacity that may be available in fair weather and schedule on the assumption of adverse weather of a certain type (typically IMC conditions, but not with thunderstorms). The decision on this depends both on the frequency of adverse weather and the economic pressures. Generally, U.S. weather is fair (recall table 1) and thus it is unlikely that the current flight scheduling approach will change.

Next, let us consider the airports. From the discussion above on linear versus nonlinear delay models, we see that airports which have demands that exceed the IMC and/or average thunderstorm effective capacity will provide much higher benefits for a given ATM improvement to reduce terminal delays. Within the U.S., this results in focusing attention on a relatively small number of airports for addressing low ceilings and visibility or heavy fog delays and the high thunderstorm frequency/high traffic airports for addressing convective weather delays.

Next, let us consider what the options are in various types of weather.

4.1 Low Ceilings and Visibility or Heavy Fog

These conditions are the most straightforward to address because the effective airport capacity and routes to be used are generally constant over the duration of an event. Many options (e.g., traffic automation to maximize the capacity of individual runways, parallel approach monitoring to permit independent operations on closely spaced parallel runways, the use of Instrument Landing System (ILS) or Microwave Landing System) MLS and/or Global Positioning System (GPS)) have been addressed in the literature and the benefits can be assessed easily using equation (1).

The most interesting capacity improvement option that is not well understood is the use of a wake vortex advisory system [Evans, Welch, 1991] which would permit reduced spacing on landing and/or takeoff when atmospheric conditions are such that wake vortices are not operationally significant (e.g., because they dissipate and/or are blown away). There is an urgent need for a better understanding of dissipation processes for the vortices and their relation to readily predicted meteorological conditions as well as detailed statistics of the fraction of time that favorable conditions for separation reductions occur during reduced capacity situations at the specific airports. We should also note that effective safety monitoring of reduced separations will also be important and that safety monitoring is poorly defined for monitoring of reduced take-off guidance.

The other "new" ATM area that should be fruitful in delay reduction is providing predictions of low C&V conditions that are matched to the traffic flow decision making. If the flight durations of the planes being delayed on the ground are relatively low, short-duration C&V predictions (e.g., 1-2 hours maximum) will suffice for delay reduction. This should be a much easier problem than the multi-hour prediction problem which has been emphasized historically by national weather services.

We currently do not have accurate statistics on how well traffic managers succeed today in matching traffic flow to effective capacity (to assess the potential benefits), and the weather prediction task may differ considerably between various airports [Clark, 1995]. It may also be appropriate to consider the use of probabilistic estimates of weather event duration as opposed today's "best guess" [Andrews, 1993].

It should also be noted that these short-term C&V predictions will also help significantly in the hold/divert decision making for a pilot in a holding pattern at the airport and thus enhance safety. Additionally, it may be beneficial to develop automation aids to assist in holding pattern control.

4.2 Convective Weather (e.g., Thunderstorms)

Assessing the ability to reduce delays due to convective weather is much more challenging than the case of low ceilings and visibility or heavy fog because of the very high variability in conditions during a thunderstorm event. The meteorological conditions that give rise to this weather often seem "chaotic" in that what appear to be quite similar conditions prior to the start of a weather event can yield storm patterns that have quite different impacts on effective capacity. Providing short-term forecasts with high specificity (e.g., "an impenetrable thunderstorm will be located at the end of a given runway in one hour") that would greatly assist in traffic capacity estimation will be extremely difficult to achieve in many situations.

Nevertheless, the initial experience with the ITWS described above has demonstrated that delay reductions can be achieved by providing highly accurate information on the current weather situation and effective precipitation estimates with prediction times of 10-20 minutes. Work is underway to extend this to approximately 30-40 minutes lead time.

Reducing thunderstorm delays is an important ATM research area in the U.S. since thunderstorms result in approximately 50 percent of the terminal weather delays. Potential areas for new ATM research include:

1. How to couple traffic automation approaches to "fuzzy knowledge" of the future weather and the pilot reaction. Our preliminary studies of pilot deviation decisions around thunderstorms have shown that a statistical result (i.e., "X percent of the pilots will seek to deviate around the cell which is 10 nmi east of the airport") may be the best that can be achieved. Most automation systems assume perfect knowledge of the future weather and the pilot reaction to that weather;

2. Estimating the "average" effective terminal capacity with current and improved weather information systems (e.g., the ITWS) to use in traffic flow planning; and

3. Understanding the extent to which current airspace structures such as TRACON/enroute transition are a major constraint in reducing terminal delays.

The initial ITWS experience has been that safety has been increased while delays have been reduced at the airports due to improved wind shear detection/prediction capability and because traffic managers could plan route usage to minimize situations in which pilots would request deviations. However, it will be necessary to continue monitoring operations with systems that facilitate more operations during thunderstorms to ensure that safety margins are not being adversely impacted by the greater flow rates.

5. Conclusions and Suggestions for Further Research

In this paper we have discussed how the efficacy of ATM options for reducing delays due to adverse terminal weather can be addressed in the context of two rather different models for delay. The preliminary work presented above suggests that the efficiency of terminal operations is a fruitful area for additional in-depth study. However, there are several topics that also warrant study that have not been discussed above.

5.1 The Benefits of Long-Term (e.g., Greater than 2-3 Hours) Predictions

The bulk of the air carrier flights into U.S. airports are less than 2-3 hours. This suggests that short-term predictions will generally suffice to effectively adjust the flow into an airport or the adjacent enroute sector to take advantage of the available effective capacity. So, what are the benefits for longer term forecasts beyond the very small number of airports that are dominated by long range flights?

For air carrier operations, the benefit would appear to lie in more nearly optimum planning for a day. How much is this worth? No analytical models for this have been reported to date. Addressing this would appear to require interaction with airline operations planners to a much greater extent than has been the case for the published studies to date.

For general aviation (GA) pilots, especially those restricted to VMC operations, longer term predictions (e.g., 1-7 days) clearly help with trip planning. How can we measure the delay reduction benefit for these GA operations? There also will be a related safety benefit associated with helping the GA pilot avoid flying in deteriorating conditions. A major challenge in developing a model for these benefits is that GA flight planning is accomplished by many independent individuals using a wide variety of data sources to accomplish their planning.

5.2 *"Free Flight" and Thunderstorms*

A basic tenet of "free flight" has been to return decision making on the route of a given flight to the pilots/airlines as much as possible. Thunderstorms cause rapidly time-varying changes in routes and are not easily monitored with airborne weather radars in the terminal area by busy pilots. Historically, the terminal weather information available on the ground was not as accurate as that aboard the plane in many cases. However, the ground capability will be increasing dramatically with the advent of weather information systems such as ITWS that provide color weather graphics displays for the ground elements of figure 1.

Hence, "free flight" must consider that there may need to be a major change in the relative roles of pilots and air traffic personnel when thunderstorms are present and that the physical region in which this transition occurs may need to be weather dependent. Addressing this will require very careful concept development and validation, and it may be necessary to consider the accelerated implementation of cockpit weather graphical services.

5.3 *Terminal Delay: A "General" Solution vs. Many Individual Solutions*

Historically, people have proposed general ATM solutions for delay but then found that these general solutions have many problems applied to specific airports. We have found that the delay causing terminal weather phenomena can differ considerably between the various high-traffic airports which have the greatest potential for producing high delay reductions for relatively modest ATM investments. Additionally, there are many airport-specific elements which need to be considered when considering the applicability of a given ATM option (e.g., the efficacy of some terminal automation options).

This suggests an approach whereby very specific assessments are done of operations during delay-causing weather events at individual airports to determine what ATM investment options will in fact be most likely to achieve significant delay reductions. Having accomplished this for a number of airports, one would then ascertain what ATM solution elements constitute a "general" solution.

This has been accomplished somewhat for options to improve effective capacity during low C&V events (e.g., studies of parallel approach monitoring and airport layout changes). However, it has yet to be accomplished for the other key ATM options discussed above.

References

[1] Andrews, J., 1993: Impact of weather event uncertainty upon an optimum ground-holding strategy, *Air Traffic Control Quarterly*, 1, 59-84.

[2] Boswell, S.B., 1997: *Analysis of downstream impacts of air traffic delay*, Report ATC-257, M.I.T. Lincoln Laboratory, 1997.

[3] Clark, D., 1995: *Characterizing the cause of low ceiling and visibility at U.S. airports*, Sixth Conference on the Aviation Weather System, American Meteorological Society, 325-330.

[4] DeArmon, J.S., 1992: *Analysis and research for traffic flow management*, Proc. of 37th Annual Conference of Air Traffic Control Association, Atlantic City, NJ, Air Traffic Control Association, 423-429.

[5] Evans, J.E. and Welch, J.D. 1991: *Role of FAA/NWS terminal weather sensors and terminal air traffic automation in providing a vortex advisory service*, FAA International Wake Vortex Symposium, Session on Operational Considerations, October 29-31, 1991, Washington, DC.

[6] Evans, J. and Clark, D., 1993: *Assessment of the benefits for improved terminal weather information*, Fifth Conference on Aviation Weather Systems, American Meteorological Society, Vienna, VA USA, 414-416.

[7] Evans, J.E., 1995: *Measuring the economic benefit of aviation meteorological products*, 14th Conference on Weather Analysis and Forecasting, American Meteorological Society, Paper J-2.

[8] Frolow, I., J. Sinott, and A. Wong, 1989: *National airspace system analysis capability: a status report after one year*, Proc. of 34th Annual Conference of Air Traffic Control Association, Boston, MA, Air Traffic Control Association.

[9] Hartman, B., 1993: The future of head-up guidance, *IEEE Aerospace and Electronic System Magazine*, 8, 31-33.

[10] Jesuroga, R., 1993a: *Using ATMS weather products for air traffic strategic planning*, Fifth Conference on Aviation Weather Systems, American Meteorological Society, Vienna, VA USA.

[11] Jesuroga, R., R. Wright, B. Campbell, 1993b: *Interactive and functional capability of the FAA's advanced traffic management system aircraft situation display*. Preprints. 9th International Conference on Interactive Information and Processing Systems, Anaheim, CA. 17-22 January 1993, American Meteorological Society, 126-128.

[12] Weber, M. E., M. Wolfson, D. Clark, S. Troxel, A. Madiwale, and J. Andrews, 1991: *Weather information requirements for terminal air traffic control automation*, Proc. Fourth International Conference of Aviation Weather Systems, Paris, France, June 28, 1991, American Meteorological Society, Boston.